NEUTRINO ASTROPHYSICS

Documents on Modern Physics

Edited by

ELLIOTT W. MONTROLL, Institute of Defense Analyses
GEORGE H. VINEYARD, Brookhaven National Laboratory
MAURICE LÉVY, Université de Paris

A. H. COTTRELL, Theory of Crystal Dislocations
A. ABRAGAM, L'Effet Mössbauer
A. B. PIPPARD, The Dynamics of Conduction Electrons
K. G. BUDDEN, Lectures on Magnetoionic Theory
S. CHAPMAN, Solar Plasma, Geomagnetism and Aurora
R. H. DICKE, The Theoretical Significance of Experimental Relativity
J. A. WHEELER, Geometrodynamics and the Issue of the Final State
BRYCE S. DEWITT, Dynamical Theory of Groups and Fields
J. W. CHAMBERLAIN, Motion of Charged Particles in the Earth's Magnetic Field
HONG-YEE CHIU, Neutrino Astrophysics
JOHN G. KIRKWOOD, Selected Topics in Statistical Mechanics
JOHN G. KIRKWOOD, Macromolecules
JOHN G. KIRKWOOD, Theory of Liquids
JOHN G. KIRKWOOD, Theory of Solutions
JOHN G. KIRKWOOD, Proteins
JOHN G. KIRKWOOD, Quantum Statistics and Cooperative Phenomena
JOHN G. KIRKWOOD, Shock Waves
JOHN G. KIRKWOOD, Dielectrics—Intermolecular Forces—Optical Rotation

Additional volumes in preparation

Neutrino Astrophysics

HONG-YEE CHIU

NASA Institute for Space Studies

and

Columbia University

Notes by Eric H. Roffman

GORDON AND BREACH
Science Publishers
NEW YORK • LONDON • PARIS

Copyright © 1965 by Gordon and Breach, Science Publishers, Inc.
150 Fifth Avenue, New York, 10011

Library of Congress Catalog Card Number 65—27511

These lectures were originally presented at the 1963 Brandeis Summer Institute in Theoretical Physics, and published in *Lectures on Astrophysics and Weak Interactions,* Copyright © 1964 by Brandeis University.

Editorial offices for Great Britain and Europe:
GORDON AND BREACH, SCIENCE PUBLISHERS LTD.
171 Strand, London, W.C.2, England

Distributed in the United Kingdom by:
Blackie & Son, Ltd.
5 Fitzhardinge Street, London, W.1, England

Printed in the United States of America

EDITORS' PREFACE

Seventy years ago when the fraternity of physicists was smaller than the audience at a weekly physics colloquium in a major university a J. Willard Gibbs could, after ten years of thought, summarize his ideas on a subject in a few monumental papers or in a classic treatise. His competition did not intimidate him into a muddle correspondence with his favorite editor nor did it occur to his colleagues that their own progress was retarded by his leisurely publication schedule.

Today the dramatic phase of a new branch of physics spans less than a decade and subsides before the definitive treatise is published. Moreover modern physics is an extremely interconnected discipline and the busy practitioner of one of its branches must be kept aware of breakthroughs in other areas. An expository literature which is clear and timely is needed to relieve him of the burden of wading through tentative and hastily written papers scattered in many journals.

To this end we have undertaken the editing of a new series, entitled *Documents on Modern Physics,* which will make available selected reviews, lecture notes, conference proceedings, and important collections of papers in branches of physics of special current interest. Complete coverage of a field will not be a primary aim. Rather, we will emphasize readability, speed of publication, and importance to students and research workers. The books will appear in low-cost paper covered editions, as well as in cloth covers. The scope will be broad, the style informal.

From time to time, older branches of physics come alive again, and forgotten writings acquire relevance to recent developments. We expect to make a number of such works available by including them in this series along with new works.

Elliott Montroll
George H. Vineyard
Maurice Lévy

CONTENTS

1. Introduction 1

2. General Properties of a Star 2
   2.1 Gravitational Binding and Equilibrium 2
       (a) Binding 2
       (b) Hydrostatic Equilibrium 3
       (c) Hydroinstability 5
       (d) Virial Theorem 8
   2.2 Limits on Stellar Size 9
       (a) The Minimum Mass of a Star 11
       (b) The Maximum Mass of a Star 13
       (c) Maximum Mass of Cold Matter 14
   2.3 Lifetime and Stellar Luminosity 17

3. Equations of Stellar Structure 22
   3.1 Radiative Equilibrium 22
   3.2 Convection 25

4. Nuclear Reactions 29
   4.1 Theory 29
   4.2 Specific Reactions 34
       (a) The Proton-Proton Reactions 34
       (b) Light Element Synthesis 34
       (c) Carbon Cycle 36
       (d) Synthesis of Carbon 36
       (e) α Process; Carbon, Oxygen Burning 37
   4.3 Pycnonuclear Reactions 38

5. The Hertzsprung-Russell Diagram 41

6. Neutrino Processes ... 47
   6.1 Neutrino Processes ... 47
      (a) The URCA Process ... 47
      (b) Neutrino Bremsstrahlung ... 49
      (c) Photo-Neutrino Process ... 51
      (d) The Pair Annihilation Process ... 54
      (e) Plasma Process ... 58
      (f) Summary ... 59
   6.2 Stellar Collapse ... 60
      (a) Neutrino Process ... 60
      (b) Fe-He Phase Transition ... 65

7. Introduction to General Relativity ... 68

8. Neutron Stars ... 75
   8.1 Schwarzschild Exterior Solution ... 75
   8.2 Schwarzschild Interior Solution;
      The Equations of a Neutron Star ... 81
   8.3 The Equations of State ... 85
      (a) Fermi Gas ... 85
      (b) Hyperon Gas ... 86
      (c) Incompressible Fluid ... 88
   8.4 Discussion ... 92
   8.5 Observation of Neutron Stars ... 93

9. A Word about Cosmology ... 95

10. Summary ... 97

Appendix ... 99
   A. Rotation ... 99
   B. Determination of the Temperature of a Star ... 102

Bibliography ... 103

    Except when there is a footnote on the same page, an asterisk in the text refers to a reference in the bibliography beginning on page 103. The bibliography also contains some general references not explicitly cited in the text.

1. INTRODUCTION

Although the title of the present lectures is "Neutrino Astrophysics," some anti-coincidence of the material covered in this course and the title is purely intentional. After I had given the title to Dr. Goldstein, I realized what a mistake it would be if I merely talked on neutrinos and supernovae.

It was remarked by a prominent physicist that half of a research problem in physics is to ask the relevant question. The other half is to find the relevant answer. In order that relevant questions be asked, it is necessary to have at least a general idea of the field. For this reason, I intend to survey the general field of astrophysics before I go into neutrino astrophysics as advertised.

One word about units. I will use the Gaussian system for electromagnetism, cgs for all other units; when nuclear transitions occur $\hbar = c = 1$. For statistical physics, $\hbar$ and c will be retained. There are also some units peculiar to astrophysics. Among these are:

$M_\odot$ = solar mass = 2×10^{33} grams

$L_\odot$ = solar energy output = solar luminosity = 4×10^{33} ergs/sec.

$R_\odot$ = solar radius = 6.95×10^{10} cm.

A.U. = radius of earth orbit = "astronomical unit" = 1.5×10^{13} cm.

Light year (l.y.) = c x 1 year $\approx 10^{18}$ cm.

Parsec = parallax of one second of arc
= 3.26 light years.

2. GENERAL PROPERTIES OF A STAR

2.1 Gravitational Binding and Equilibrium

A star is a gravitationally bound object in mechanical equilibrium. (We will ignore rotational effects.) Its structure is described by a set of differential equations. We shall describe how a star achieves equilibrium first by a variational principle and then just by considering the balance of forces.

The energy of a system in stable equilibrium must be a minimum. Let the total energy of the star be E_T, then

$$\delta E_T = 0 \qquad \text{condition for equilibrium;} \qquad (2.1)$$

$$\delta^2 E_T \geqslant 0 \qquad \text{condition for stable equilibrium.} \qquad (2.2)$$

Furthermore, the total energy must be less than zero for the object to be bound gravitationally:

$$E_T < 0 \qquad (2.3)$$

(a) Binding

Given a spherical object of mass M, mean density $\rho \equiv M/\frac{4}{3}\pi R^3$ and radius R, its total gravitational energy, W_g, is

$$W_g = -\int_0^M \frac{m\,dm}{r} \approx -\frac{GM^2}{2R} = -\frac{M}{2} G\rho \left(\frac{4}{3}\pi R^2\right) ; \qquad (2.4)$$

$m = m(r)$ is the mass contained within the radius r. For uniform density

$$m(r) = \left(\frac{4}{3}\pi r^3\right)\rho . \qquad (2.5)$$

NEUTRINO ASTROPHYSICS

Its thermodynamic energy may be characterized by a temperature T. The thermal energy, E_{th}, is roughly

$$E_{th} \approx 3 M R_g T. \tag{2.6}$$

The total energy E_T is

$$E_T = W_g + E_{th} = M \left\{ 3 R_g T - \frac{2\pi}{3} G R^2 \rho \right\}$$

$$= M \left\{ 3 R_g T - \frac{M G}{2 R} \right\} \tag{2.7}$$

$$= M \left\{ 3 R_g T - \frac{1}{2} G \left[\frac{4\pi}{3} \rho \right]^{1/3} M^{2/3} \right\}.$$

Given a value of ρ and of T, one can always find a value of M such that it is gravitationally bound ($E_T < 0$).

The temperature of interstellar gas is $\sim 10^0$ K and the particle density is ~ 1 particle/cc. Then the radius of a mass of gas sufficient for binding is ~ 100 l.y.

(b) <u>Hydrostatic Equilibrium</u>

The stability of a star is determined by the conditions:

$$\delta^1 E_T = 0, \tag{2.8}$$

$$\delta^2 E_T \geq 0. \tag{2.9}$$

We will neglect rotation in the following discussion. It will turn out not to be an important effect, and we shall return to it later.

The variable m(r), the mass within a radius r, is given by

$$m(r) = \int_0^r 4\pi r^2 \rho(r) \, dr \tag{2.10}$$

The volume V(r) of the sphere of radius r is

$$V(r) = \frac{4\pi}{3} r^3. \tag{2.11}$$

We also use the specific volume v(r), the volume occupied by one gram of matter:

$$v(r) \equiv \frac{1}{\rho(r)} = \frac{dV(r)}{dM(r)} \qquad (2.12)$$

Now we must compute the variation of E_T:

$$\delta E_T = \delta \left[-\int_0^M \frac{Gm(r)\, dm(r)}{r} + \int_0^M \epsilon_{th}(r)\, dm(r) \right]; \qquad (2.13)$$

$\epsilon_{th}(r)$ is the energy/gm at the radius r.

$$\delta E_T = \int_0^M \delta r \frac{m\, dm}{r^2} + \int_0^M \delta \epsilon_{th}\, dm . \qquad (2.14)$$

Expressing ϵ_{th} as a function of the specific volume at r, and considering adiabatic changes of ϵ_{th}, we may write

$$\delta \epsilon_{th} = \left(\frac{\partial \epsilon(v)}{\partial v} \right)_{ad.} \delta v . \qquad (2.15)$$

Using $v = dV/dm$ and the fact that δ commutes with d,

$$\delta \epsilon_{th} = \left(\frac{\partial \epsilon(v)}{\partial v} \right)_{ad.} \frac{d(\delta V)}{dm} , \qquad (2.16)$$

$$\delta V = 4\pi r^2 \delta r . \qquad (2.17)$$

From thermodynamics, we have

$$P dv + d\epsilon = T ds . \qquad (2.18)$$

For an adiabatic process, ds = 0. Hence

$$P = -\left(\frac{d\varepsilon}{dv}\right)_{ad}. \tag{2.19}$$

Combining terms, we have

$$\delta E_T = \int_0^M \delta r \left[\frac{Gm}{r^2} + 4\pi r^2 \frac{dP}{dm}\right] dm = 0. \tag{2.20}$$

For the star to be stable against adiabatic variation in which $\delta r = 0$ at $m = 0$, and for which $P = 0$ at the boundary of the star, this condition may be written

$$\frac{Gm}{r^2} = -\frac{dP}{dm}(4\pi r^2). \tag{2.21}$$

The differential form of Eq. (2.10) is

$$\frac{dm}{dr} = 4\pi r^2 \rho(r). \tag{2.22}$$

Using this, Eq. (2.21) becomes

$$\frac{dP}{dr} = -\frac{\rho(r)Gm(r)}{r^2}. \tag{2.23}$$

The term on the right in Eq. (2.23) is the gravitational force on a cubic centimeter of matter at the radius r. The term on the left represents the change per cm of the force per sq cm. It is the hydrostatic pressure gradient. Thus the result of the variational principle calculation is the result familiar from hydrodynamics. The force is balanced by the pressure gradient.

(c) <u>Hydrostatic Instability</u>*

We now pass on to the more interesting question of determining the second variation. The equilibrium determined by $\delta E_T = 0$ may be stable or unstable.

First we calculate the second variation of ϵ_{th}:

$$\tfrac{1}{2}\delta^2 \epsilon_{th} = \tfrac{1}{2}\delta\left[\left(\frac{d\epsilon}{dv}\right)_{ad.} \delta v\right]$$

$$= \tfrac{1}{2}\left[\left(\frac{\partial^2 \epsilon}{\partial v^2}\right)_{ad.} (\delta v)^2 + \left(\frac{\partial \epsilon}{\partial v}\right)_{ad.} \delta^2 v\right]. \tag{2.24}$$

Using the relations $v = \frac{dV}{dm}, \left(\frac{\partial \epsilon}{\partial v}\right)_{ad} = -P$

we find

$$-\tfrac{1}{2}\left(\frac{\partial P}{\partial v}\right)_{ad.} (\delta v)^2 - P\frac{d}{dm}\left(4\pi r (\delta r)^2\right) = \tfrac{1}{2}\delta^2 \epsilon_{th}. \tag{2.25}$$

Since

$$-\frac{v}{P}\left(\frac{dP}{dv}\right)_{ad} \equiv \gamma \tag{2.26}$$

is the adiabatic exponent of a gas,

$$\tfrac{1}{2}\delta^2 \epsilon_{th} = \tfrac{1}{2}\frac{P}{v}\gamma\left[\frac{d}{dm}\delta V\right]^2 - P\frac{d}{dm}\left[4\pi r (\delta r)^2\right]. \tag{2.27}$$

From this result for $\delta^2 \epsilon_{th}$ we find, for $\delta^2 E_T$,

$$\delta^2 E_T = \int_0^M dm\left\{-\frac{Gm}{v^3}(\delta r)^2\right\}$$

$$+ \int_0^M \left\{\tfrac{1}{2}\frac{P\gamma}{v}\left[\frac{d(\delta V)}{dm}\right]^2 - P\frac{d}{dm}(4\pi r(\delta r)^2)\right\} dm. \tag{2.28}$$

Integrating the third term by parts and utilizing Eq. (2.20) and the relation $V = \tfrac{4}{3}\pi r^3$ to give $(\delta r)^2 = (\delta V)^2/(12\pi rV)$, we have, finally,

$$\delta^2 E_T = \int_0^M dm \left[\frac{2}{3} \frac{dP}{dm} \frac{(\delta V)^2}{V} + \frac{P\gamma}{2} \frac{dm}{dv} \left(\frac{d(\delta V)}{dm}\right)^2 \right]. \qquad (2.29)$$

For the stability of the star, this expression for $\delta^2 E_T$ must satisfy $\delta^2 E_T \geqslant 0$ for any possible perturbation δV.

Consider a possible variation of the star having the form $\delta V = \alpha V$. We change variables from m to V, so that

$$\frac{dP}{dm} dm = \frac{dP}{dV} dV \qquad (2.30)$$

$$\left(\frac{d(\delta V)}{dm}\right)^2 \frac{dm}{dV} dm = \left(\frac{d(\delta V)}{dV}\right)^2 dV. \qquad (2.31)$$

Now the equation for $\delta^2 E_T$ may be written in terms of the new variable:

$$\delta^2 E_T = \frac{2}{3} \int_0^V dV \left\{ \frac{1}{V} \frac{dP}{dV} (\delta V)^2 + \frac{3}{4} \gamma P \left(\frac{d(\delta V)}{dV}\right)^2 \right\}. \qquad (2.32)$$

Now we substitute $\delta V = \alpha V$ (a change of scale) and then integrate by parts,

$$\delta^2 E_T = \frac{2}{3} \alpha^2 \int_0^V dV \left\{ \frac{dP}{dV} V + \frac{3}{4} \gamma P \right\}$$

$$= \frac{2}{3} \alpha^2 \int_0^V dV \left[\gamma - \frac{4}{3} \right] P > 0. \qquad (2.33)$$

The criterion for stability is that $[\gamma - 4/3]P$ integrated over the star be greater than zero, i.e., γ must be $\geqslant 4/3$ over much of the star. For a classical ideal gas $\gamma = c_p/c_v = 5/3$.

The adiabatic exponent is defined as

$$-\frac{v}{P}\left(\frac{dP}{dv}\right)_{ad} \equiv \gamma. \qquad (2.26)$$

In general P, v, ϵ, may be functions of various thermodynamic variables, although in general it is adequate to admit only two. Let x and y be independent thermodynamic variables. For an adiabatic change,

$$0 = Pdv + d\epsilon = (Pv_x + \epsilon_x)dx + (Pv_y + \epsilon_y)dy , \qquad (2.34)$$

where

$$v_x \equiv \frac{\partial v}{\partial x} . \qquad (2.35)$$

We also may write

$$dP = P_x dx + P_y dy , \qquad (2.36)$$

$$dv = v_x dx + v_y dy . \qquad (2.37)$$

These equations, combined with Eq. (3.34) give

$$\gamma = -\frac{v}{P}\frac{dP}{dv} = -\frac{v}{P}\frac{(P_x \epsilon_y - P_y \epsilon_x) + P(P_x v_y - P_y v_x)}{v_x \epsilon_y - v_y \epsilon_x} . \qquad (2.38)$$

For a classical gas $P = \rho RT$, $\epsilon = (3/2)\rho RT$, $\gamma = 5/3$.

(d) <u>Virial Theorem</u>

The equation of mechanical equilibrium (2.23) can be written

$$VdP = -\rho V \frac{Gm}{r^2} dr \qquad (2.39)$$

$$= -\rho \frac{Gm}{r} \frac{4}{3}\pi r^2 dr .$$

Integrating,

$$\int_0^M VdP = -\frac{1}{3}\int_0^M \frac{Gm\,dm}{r} = +\frac{1}{3} W_g , \qquad (2.40)$$

NEUTRINO ASTROPHYSICS

and integrating by parts,

$$3\int_0^V P\, dV = -W_g . \tag{2.41}$$

Equation (2.41) is a general form of the Virial Theorem. The form of the equation for a perfect gas may be obtained by substituting

$$P = \rho RT, \quad \epsilon = 3/2 \rho RT, \quad P = 2/3 \epsilon, \tag{2.42}$$

$$-2\int \epsilon_{th} dV = W_g \tag{2.43}$$

In this case

$$E_T = W_g + E_{th} = -E_{th} = -\int \epsilon_{th} dV \tag{2.44}$$

For radiant energy, $P = (1/3)\epsilon$. We see that

$$\int \epsilon\, dV = -W_g$$

In this case the total energy $E_T = 0$. Since for gravitational binding $E_T < 0$ we see that in this case there is stability trouble.

2.2 Limits on Stellar Size

We are now going to apply our theoretical discussion of stability and the Virial Theorem to try to understand why astrophysical quantities have a particular range of values.

On the next page we list the range of observed values of certain stellar quantities, and for each quantity give the value for the sun.

	Stars	Sun
Mass	2×10^{32} —— 2×10^{35} g	2×10^{33} g
Energy output rate	up to 4×10^{38} erg/sec	4×10^{33} erg/sec
Mean density	$10^{-6} - 10^{7}$ g/cm^3	1 g/cm^3
Radius	$10^{7} - 10^{13}$ cm	7×10^{10} cm
Age	up to 2.4×10^{10} years	4.5×10^{9} years
Surface temperature	$10^{3} - 5 \times 10^{4}$ °K	6,000 °K
Average temperature	$\approx 10^{7}$ °K	7×10^{6} °K
Central temperature	$1.3 - 3 \times 10^{7}$ °K	1.5×10^{7} °K

We will use the following units:

length: $\hbar/m_e c = 0.4 \times 10^{-10}$ cm;

time: $[\hbar/m_e c]1/c = 10^{-21}$ sec;

energy: $m_e c^2 = 0.5$ MeV.

Now we derive some convenient units. Let R_a be the side of a cube in which, on the average, one proton is found. Then the density can be written

$$\rho = m_p/R_a^3, \quad \text{where} \quad m_p = \text{proton mass}.$$

We write $r_a = R_a$ in units of $\hbar/m_e c$, that is

$$r_a \equiv R_a/\hbar/m_e c. \tag{2.46}$$

Then

$$\rho = m_p(m_e c/\hbar)^3/r_a^3 = \frac{2.61 \times 10^7}{r_a^3} \text{ g/cm}^3. \tag{2.47}$$

Let

$$\alpha_G \equiv \frac{Gm_p^2}{\hbar c} = 0.5 \times 10^{-38}; \tag{2.48}$$

NEUTRINO ASTROPHYSICS

α_G is like a "fine structure constant for gravitation."
Let
$$N_o \equiv \alpha_G^{-3/2} = 2 \times 10^{57}. \qquad (2.49)$$

N_o is a particle number and is roughly equal to $2 N_\odot$, where $N_\odot$ is the number of nucleons in the sun.
Mass may be written

$$M = N m_p = \frac{4\pi}{3} \rho R^3, \qquad (2.50)$$

and the radius can be expressed as

$$R = \sqrt[3]{\frac{3}{4\pi}} N^{1/3} \left(\frac{\hbar}{m_e c} r_a \right) \approx \frac{\hbar}{m_e c} N^{1/3} r_a . \qquad (2.51)$$

When interstellar gas condenses to form stars, it must contract first until sufficient gravitational energy is released to raise its temperature to a degree suitable for hydrogen nuclear reactions to proceed. In order that this be possible, a star must be massive enough. There is also an upper limit for the mass of a star due to the fact that radiation pressure must not dominate. Hence we want to examine in detail the following:

1 - The existence of a lower mass limit for a star;

2 - The existence of an upper limit for the mass of a star;

3 - The existence of an upper limit to the mass of cold matter;

4 - Such things as the lifetimes and liminosities of stars.

All this will be done essentially by dimensional analysis, and without solving the differential equations.

(a) <u>The Minimum Mass of a Star</u>

The average gravitational energy per proton is

$$\epsilon_{gr} = \frac{G m_p (m_p N)}{R} = \alpha_G N^{2/3} \frac{1}{r_a} m_e c^2 . \qquad (2.52)$$

Measuring the thermal energy kT and ϵ_{gr} in units of $m_e c^2$, they must be, by the Virial Theorem, of the same order of magnitude. Hence

$$\epsilon_{gr}/m_e c^2 \simeq kT/m_e c^2 \simeq \alpha_G N^{2/3} \frac{1}{r_a} . \qquad (2.53)$$

By Eq. (2.49), we write

$$kT/m_e c^2 = (N/N_o)^{2/3} \frac{1}{r_a} = (N/2N_\odot)^{2/3} \frac{1}{r_a} . \qquad (2.54)$$

Equation (2.54) shows the appropriateness of the definition in Eq. (2.49).

We are considering fermions, hence the exclusion principle limits the number of particles to one per cubic de Broglie wavelength.

The wavelength of a particle is

$$\lambda = \hbar/p \qquad (2.55)$$

and the requirement is

$$\hbar/p = \lambda \lesssim r_a \hbar/m_e c . \qquad (2.56)$$

In non-relativistic approximation

$$p^2/2m = kT , \qquad (2.57)$$

so

$$\lambda = \hbar/(2m_e kT)^{1/2} . \qquad (2.58)$$

Define

$$\Lambda \equiv \lambda/\hbar/m_e c \lesssim r_a . \qquad (2.59)$$

Then with

$$T_o \equiv m_e c/k = 6 \times 10^9 \, ^\circ K , \qquad (2.60)$$

algebra shows

$$r_a \gtrsim \Lambda = (T_o/2T)^{1/2} \qquad (2.61)$$

From Eqs. (2.54) and (2.61) we have, dropping the factor of 2,

$$r_a = (T/T_o)^{-1}(N/N_o)^{2/3} \gtrsim (T_o/T)^{1/2} . \qquad (2.62)$$

Hence

$$N/N_o \gtrsim (T/T_o)^{3/4} . \qquad (2.63)$$

Given a star of a certain size, the exclusion principle limits the temperature to which it can rise by means of gravitational contraction, as we shall see later, since a certain temperature is needed to ignite nuclear reactions. This puts a requirement on the minimum mass needed for a gas to contract to a star.

In order for a nuclear reaction to occur, T must be greater than about 10^7 °K or $(2 \times 10^{-3})T_o$. Then from Eq. (2.63), $M \gtrsim 0.02 \, M_\odot$ for nuclear reactions to be ignited by a contracting gas mass.

For thermal energy to be great enough to ionize hydrogen,

$$kT \approx \alpha^2 mc^2 \approx 27 \text{ eV}. \qquad (2.64)$$

Then $M \gtrsim 0.001 \, M_\odot$.

There is some observational support for these limits. No star has been observed with $M < 0.05 \, M_\odot$, and Jupiter, a cold body, has mass $M \approx 0.96 \times 10^{-3} \, M_\odot$. This makes Jupiter one of the largest cold bodies that exists. The internal structure of Jupiter is very interesting. For instance, some people believe the center is degenerate.

(b) The Maximum Mass of a Star

The exclusion principle put lower limits on the mass of stars. We will now look at the upper limits. Here the criterion is stability.

In Section 2.1 (c) we derived the stability condition

$$\int_o^V [\gamma - 4/3] P dV \gtrsim 0 . \qquad (2.65)$$

For radiation pressure, $\gamma = 4/3$. Consequently the gas pressure must not be less than the radiation pressure. The gas energy per particle is

$$E_g = kT \, [m_e c/\hbar)/r_a]^3 . \qquad (2.66)$$

The radiation energy density is roughly equal to

$$E_R = kT/\lambda^3 ; \qquad (2.67)$$

λ is the wavelength of the light.
E_g depends on density, while E_R does not.
Let

$$\Lambda \equiv \lambda/\hbar/m_e c = (T_o/2T)^{1/2} . \qquad (2.68)$$

Then

$$E_{Rad}/E_{gas} = (r_a/\Lambda)^3 \qquad (2.69)$$

and by Eq. (2.62),

$$E_{Rad}/E_{gas} = (1/2\pi)^3 (N/N_o)^2 = (N/16 N_o)^2 = (N/30 N_\odot)^2 . \qquad (2.70)$$

If $N \gg 30 \, N_\odot$, the radiation density will dominate and the star will not be stable.

A sophisticated calculation of a model by Schwarzschild and Härm showed that when a star has a mass larger than this mass limit, it begins to oscillate with large amplitudes, presumably until enough mass is ejected to bring it within this mass limit, although no one has followed the model to the ejection of matter.

(c) <u>Maximum Mass of Cold Matter</u>

For a zero temperature classical gas there is no kinetic pressure. Zero temperature stars do exist. Hence, some other source of pressure must exist.

Bosons at zero temperature exert no pressure. However, stars can be made at zero temperature from fermions. This pressure is due to the exclusion principle. In cold matter of not too high density, the available materials are e^-, p^+, n^o.

NEUTRINO ASTROPHYSICS 15

For most stars only electrons contribute to the pressure. Protons and neutrons, having higher mass, contribute less to the Fermi pressure. The neutrons are unstable at usual white dwarf density (10^7 g/cm^3). The neutrons become stable only at an extremely high density (10^{12} g/cm^3). When the Fermi energy of the electron exceeds the neutron - proton mass difference, then the inverse beta reaction proceeds:

$$e^- + p^+ \longrightarrow n + \nu \qquad (2.71)$$

Neutron stars would have a density $\sim 10^{14}$ g/cm^3. They are very interesting objects for study, and we shall discuss them later. I believe they are the endpoints of the supernova explosion. Although no stars of this density have been found, it does not mean they do not exist. It only means that we have not sought hard enough.

For a Fermi gas at zero temperature all momentum states are occupied up to a momentum p_F; none are occupied for $p > p_F$. The average energy of the electrons, if $p_F \gg m_e c$, is

$$E = p_{mean} c ; \qquad (2.72)$$

and, in the non-relativistic case, when $p_F \ll m_e c$,

$$E = p^2/2 m_e . \qquad (2.73)$$

We may write

$$p = \hbar/\lambdabar = \hbar/(r_a \, \hbar/m_e c) = m_e c/r_a , \qquad (2.74)$$

which is the momentum written in terms of the number of Compton wavelengths between particles.

The energy formulas can be combined into the following approximate formula:

$$E_F/m_e c^2 = \frac{1}{r_a + r_a^2/2} , \qquad (2.75)$$

which is valid within about 8% over the whole range of energy, as verified by Wheeler.*

The Virial Theorem says that

$$E_F/m_e c^2 \approx E_G/m_e c^2 \tag{2.76}$$

and as we saw in Eqs. (2.52) - (2.54),

$$\frac{1}{r_a} \frac{1}{1+r_a/2} = \frac{E_F}{m_e c^2} = \left(\frac{N}{N_o}\right)^{2/3} \frac{1}{r_a} \tag{2.77}$$

or:

$$\frac{1}{1+r_a/2} = (N/N_o)^{2/3}. \tag{2.77a}$$

We see that for $N < N_o$ Eq. (2.77a) can be satisfied for some positive value of r_a. For $N > N_o$, no positive solution for r_a exists, hence there is no equilibrium configuration. Thus for masses much greater than $2\,M_\odot$, the Fermi pressure is not sufficient to support the star.

A detailed analysis of the stability of cold Fermi gas stars is given by Chandrasekhar[*], who obtained the result

$$M \lesssim 1.4\,M_\odot, \quad \text{He star;} \tag{2.78}$$

$$M \lesssim 5.6(Z/A)^2\,M_\odot \quad \text{in general}$$

For $M \sim 1\,M_\odot$, one can verify from Eq. (2.77) that $r_a \approx 1$. Consequently, the spacing between particles is $\approx$ a Compton wavelength. The density is

$$\rho \sim m_p/(\hbar/m_e c)^3. \tag{2.79}$$

Equations (2.78) and (2.79) indicate the radii of these stars are $10^8 - 10^9$ cm or about the same as the radius of the earth. These stars are called white dwarfs.

Because of the large mass and small radius, they have a large gravitational red shift -- about 30 A for the yellow line. The nearest white dwarf is the companion of Sirius A, which was predicted because of periodic motion of Sirius A, and first seen by Alven Clark in 1893 while testing the telescope at Yerkes Observatory. The red shift of 40 Eridanus B has been measured and is consistent with the relativistic predictions.

2.3 Lifetime and Stellar Luminosity

(a) Rough Estimates

Consider the following model for radiative transfer. A series of sheets, 1, ---, m at temperatures T_1, ---, T_m, each absorbing and re-radiating black-body radiation, are arranged in line (see Fig. 2.1).

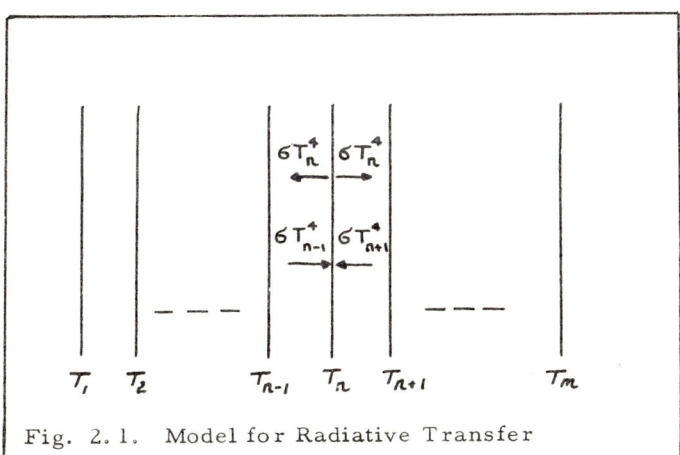

Fig. 2.1. Model for Radiative Transfer

The plate T_1 is maintained at a constant temperature. The plates are in thermal equilibrium. If a steady state exists, then the same amount of energy flows through all plates. The same amount of energy must flow from plate T_1 to plate T_2, from plate T_2 to plate T_3, etc. Plate T_1 radiates an amount of energy σT_1^4 to plate T_2 in a unit time from a unit area, and plate T_2 radiates an amount of energy σT_2^4 back to plate T_1. The net energy flow from plate T_1 to plate T_2 is thus $\sigma T_1^4 - \sigma T_2^4$ per unit time per unit area. That from plate T_2 to T_3 is $\sigma T_2^4 - \sigma T_3^4$, ... In a steady state they are all equal, thus

$$\sigma T_1^4 - \sigma T_2^4 = \sigma T_2^4 - \sigma T_3^4 = \ldots = \sigma T_m^4. \quad (2.80)$$

The solution of this set of equations, subject to the boundary condition of heat supplied at T_1 and radiated away at T_m, is

$$T_m^4 = T_1^4/m, \qquad T_{m-k}^4 = \frac{k+1}{m} T_1^4 . \qquad (2.81)$$

In other words, every time a photon is absorbed and rescattered the intensity factor goes down by one. For a mean free path of 0.1 g/cm^2, appropriate to light scattering in the sun, a mean density of 1 g/cm^3 and a radius of 10^{11} cm, the surface temperature is related to the central temperature by

$$T_s^4 = T_c^4/10^{12} = (T_c/10^3)^4 . \qquad (2.82)$$

For $T_c = 10^7 \ °K$,

$$T_s \sim 10^4 \ °K . \qquad (2.83)$$

The agreement with the observed value $T \sim 10^{3.7} \ °K$ must be regarded as satisfactory from such a crude model.

The luminosity L of a star is related to the surface temperature T_s by

$$L = 4\pi R^2 \sigma T_s^4 . \qquad (2.84)$$

Nuclear reactions are capable of producing an energy of about 8 MeV/particle, since that is the binding energy per particle of Fe^{56}, the most stable element. The total nuclear energy reserve, E_N of the sun is therefore

$$E_N = 8 \text{ MeV/proton} \times 10^{57} \text{ protons in sun} \times 1.6 \times 10^{-6} \frac{\text{ergs}}{\text{MeV}} \qquad (2.85)$$

$$= 1.4 \times 10^{52} \text{ ergs}.$$

The lifetime of a star is =(AVAILABLE ENERGY)/(LUMINOSITY). After about 1/10 the available nuclear energy is used, the luminosity vastly increases and the rest of the life, T, of the star is short. Therefore, $T = \frac{1}{10} E_N/L$. For the sun,

$$T = \frac{1.4 \times 10^{52}/10 \text{ ergs}}{4 \times 10^{33} \text{ erg/sec}} = 3 \times 10^{17} \text{ sec} \approx 10^{10} \text{ yr} . \qquad (2.86)$$

You can rest assured that only 5×10^9 yr. has passed.

NEUTRINO ASTROPHYSICS

(b) A Simple Model

The luminosity of a star is roughly

$$L = \frac{\text{total photon energy in star}}{\text{time for energy to diffuse out}} = \frac{(4\pi/3)R^3 \, a \, \langle T^4 \rangle}{t} \tag{2.87}$$

The time for a photon to diffuse is given by the random walk time

$$t = R/c \cdot R/\lambda = \text{const } R^2 \tag{2.88}$$

where λ is the mean free path.
Then

$$L \propto R\lambda \langle T^4 \rangle . \tag{2.89}$$

Since (Eqs. (2.51) and (2.62))

$$r_a = T_0/T(N/N_0)^{2/3}, \quad R = N^{1/3} r_a \, \hbar/m_e c ,$$

we see that

$$R \propto N . \tag{2.90}$$

The mean free path is

$$\lambda = 1/\sigma n , \tag{2.91}$$

where n, the electron density is

$$n = 1/(R_a)^3 = 1/r_a^3 (\hbar/m_e c)^3 \tag{2.92}$$

and the cross section is (Compton scattering)

$$\sigma = \frac{8}{3}\left(\frac{e^2}{m_e c^2}\right)^2 . \tag{2.93}$$

We define a dimensionless cross section S and time τ by

$$S = \sigma/(\hbar/m_e c)^2 = \frac{8\pi}{3}\alpha^2 \tag{2.94}$$

$$\tau = t/t_0 , \quad t_0 = (\hbar/m_e c)\frac{1}{c\alpha_G} \approx 7 \times 10^9 \, y .$$

Eliminating λ and R and r_a, we find

$$\tau = ST/T_o \qquad (2.95)$$

We can write

$$E_{photon} = \left(\frac{E_{photon}}{E_{gas}}\right) NkT \qquad (2.96)$$

and we have already evaluated the ratio (Eq. (2.70)):

$$\frac{E_{ph.}}{E_g} = \frac{1}{(2\pi)^3} (N/N_o)^2 . \qquad (2.97)$$

Putting this in, we find for L

$$L \approx \frac{1}{225}\left(\frac{N}{N_o}\right)^3 \frac{1}{S} \frac{N_o m c^2}{t_o} \qquad (2.98)$$

$$\approx \frac{1}{225}\left(\frac{N}{N_o}\right)^3 \frac{1}{S} L_o ,$$

$$L_o = N_o m_e c^2/t_o = 2.3 \times 10^{34} \text{ ergs/sec} .$$

The process which contributes to the scattering, hence the value of S, depends on stellar conditions. Roughly S varies as a certain power of (N_o/N), and in the intermediate mass range $S \sim (N_o/N)$. Hence $L \sim (N/N_o)^4$, which is in rough agreement with the observed $L \sim (N/N_o)^{3.7}$ for stars of intermediate mass (0.8 - 5 $M_\odot$).

Problem:

Given a star with the following density distribution function

$$\rho = \rho_o e^{-r^2/\alpha^2}$$

(1) Find an expression for the gravitational and the total energy. Use $P = \rho R_g T$ for the equation of state.

(2) Choose as boundary conditions $\rho_o = 100$, $M = 2 \times 10^{33}$. Find the "radius" .

(3) Assuming the sphere undergoes adiabatic homologous ($\delta r = \text{const.} \times r$) oscillation, find the period of oscillation for

$$\Gamma = \rho/P \left(\frac{dP}{d\rho}\right)_{ad.} = 5/3 .$$

Hint: $E = E_o + \frac{1}{2}\left(\frac{\delta^2 E}{\delta r^2}\right)_{ad} (\delta r)^2$. The second term may be treated as a potential which gives rise to simple harmonic motion.

3. EQUATIONS OF STELLAR STRUCTURE

In Chapter 2 we saw how general physical principles and dimensional analysis can set limits to the mass, age and luminosity of stars. We now will discuss the equations that govern the detailed stellar structure.

3.1 Radiative Equilibrium

From consideration of stellar equilibrium, we derived in Chapter 2 the requirement of pressure balance, or hydrostatic equilibrium:

$$\frac{dP}{dr} = -\rho \frac{Gm}{r^2} , \qquad (3.1)$$

$$\frac{dm}{dr} = 4\pi r^2 \rho . \qquad (3.2)$$

The equation of continuity must be satisfied by the radiation:

$$\frac{dE}{dt} + \frac{dL_r}{dr} = 0 \qquad (3.3)$$

where dE/dt is the rate of energy production per unit thickness of the shell of radius r and L_r is the energy flux out of the sphere of radius r.

If we define $\epsilon(\rho, T)$ as the energy generated (by nuclear reactions, gravitational contraction, etc.) per gram of matter, then

$$dL_r/dr = 4\pi r^2 \rho \epsilon = dE/dt . \qquad (3.4)$$

This is the first equation of radiative equilibrium. The second equation, which we derive below, is

$$L_r = 4\pi r^2 c \frac{1}{K\rho} \frac{4}{3} aT^3 \frac{dT}{dr} ; \qquad (3.5)$$

K, called the mass absorption coefficient, is defined so that $\frac{H_r}{c} K \rho$

NEUTRINO ASTROPHYSICS 23

is the momentum absorbed by matter per second per cubic centimeter from the radiation. H_r/c is the momentum flux of radiation,

$$\frac{H_r}{c} = \frac{L_r/c}{4\pi r^2} . \qquad (3.6)$$

The force due to radiation is $-dP_{rad}/dr$;

$$P_{rad} = \frac{1}{3}aT^4, \quad a = 7.56 \times 10^{-15} \frac{erg}{cm^3 deg^4} . \quad (3.7)$$

Then $F = dP/dt$ is equivalent to

$$\frac{d}{dr}(\frac{1}{3}aT^4) = \frac{L_r}{c} \frac{1}{4\pi r^2} K\rho . \qquad (3.8)$$

Finally,

$$L_r = 4\pi r^2 c \frac{1}{K\rho} \frac{4}{3} aT^3 \frac{dT}{dr} ; \qquad (3.5)$$

K_ν is the opacity at the frequency ν. K is the mean opacity, defined in the following way

$$\frac{1}{K} = \frac{\int_0^\infty \frac{1}{K_\nu} \frac{dI_\nu}{dT} d\nu}{\int_0^\infty \frac{dI_\nu}{dT} d\nu} \qquad (3.9)$$

I_ν is the intensity of radiation of frequency ν. Computing the average of $1/K_\nu$ rather than K_ν is appropriate because small values of opacity are the most important. If for some ν K_ν is very small in some region of the star, say $r_1 \leq r \leq r_2$ then light of that frequency will travel through the region without substantial attenuation. Moreover suppose light of frequency ν' is absorbed. Because of the presence of radiation of frequency ν, by the phenomenon of stimulated emission, the atom that absorbed the photon ν' is more likely to re-emit a photon of frequency ν, which then passes out of the region. An interval of frequencies $\nu_1 \leq \nu \leq \nu_2$ where K_ν is small and hence radiation can escape is called a window.

The determination of the structure of a star in radiative equilibrium now depends on evaluating ϵ and κ. The calculation of ϵ will be considered later. Let us take a brief look at κ.

The first process which can absorb photons is

$$e^- + (Z, A) + \gamma \rightarrow e^- + (Z, A), \qquad (3/10)$$

the "free-free transition." This same process in the opposite direction is bremsstrahlung.

Bound-free transitions

$$\gamma + (Z, A) \rightarrow e^- + (Z, A)^+ \qquad (3.11)$$

are more important than the bremsstrahlung process by a factor of 100 or so in normal stars.

Finally, Thomson scattering of free electrons:

$$e^- + \gamma \rightarrow e^- + \gamma . \qquad (3.11a)$$

At usual stellar energy there is negligible transfer of energy from the photon to the electron.

The cross section for the first process is

$$\sigma \approx \frac{2 e^6 Z^2}{3\sqrt{3}\, \hbar c\, m_e^2\, \nu^3} g . \qquad (3.12)$$

For the second, it is

$$\sigma \approx \frac{1}{3\pi^2 \sqrt{3}} \frac{m e^{10}}{\hbar^6 c} \frac{Z^4}{\nu^3} \frac{1}{n^5} g \qquad (3.13)$$

The quantity g is called the Gaunt factor. It is a dimensionless correction factor, which varies within ten per cent of unity as a function of n, the principal quantum number and ν, the frequency.

Opacity depends strongly on the composition. In the interior of the star hydrogen is totally ionized, as well as most of the light elements. Consequently they cannot contribute to bound-free absorption. Iron is totally ionized at an energy

$$E_{ionization} = Z^2 \alpha^2 mc^2 \qquad (3.14)$$

$$\approx 3000 \times 27 \text{ eV}$$

$$= 80 \text{ KeV (corresponds to } \sim 10^9 \text{ °K)}.$$

At lower temperatures the metals contribute at their absorption edges. The opacity is really a quite complicated function of frequency. Russell proposed a particular composition of metals for a star. The abundances of O : Na + Mg : Fe : Sc : K + Ca are postulated to be in the ratio by weight of 8:4:2:1:1.

For this standard mixture the variation of opacity with frequency is shown in Fig. 3.1.

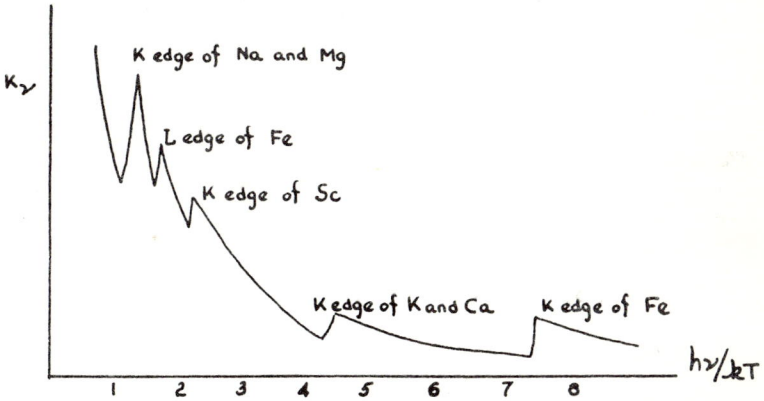

Fig. 3.1. Spectral Opacity of Standard Mixture of Metals. (Source: Chandrasekhar, Stellar Structure, p. 266.)

The scale of k_ν depends on the total proportion of metals. A typical population I star (like our sun) may have 60% by weight hydrogen and from 3 to 10% metals. The rest is helium. A population II star may have ten times less metal than that of a population I star.

Thomson scattering is the simplest process, with the familiar cross section

$$\sigma = \frac{8\pi}{3} r_o^2 , \qquad (3.15)$$

where $r_o = e^2/mc^2$ is the classical electron radius.

3.2 Convection

In daily life we know hot air rises, carrying energy with it, and cold air falls. In a star, also, energy can be carried by macroscopic motion of gases.

Gas which is not at a uniform temperature will normally come to equilibrium with the cold air at the bottom, and the hot air on top, if gravity alone acts. In a star the situation is the opposite; the hot material is at the center. Under what conditions will this hot material rise? When it does rise, it will carry thermal energy with it, and the equation of radiative energy transport will be replaced by the equation of convective energy transport in determining stellar structure. This is because convection is a very fast process.

Suppose a small element of material rises adiabatically from a region of density ρ_o to a region of density ρ' (see Figure 3.2). Since it undergoes an adiabatic change it will have a new density $\rho = \rho_o(P/P_o)^{1/\gamma}$, where P and P_o are the pressures at the final and initial spots. Note that the pressure inside must always be equal to the pressure outside. The density inside, ρ, is determined by the adiabatic equation from the density ρ_o, and is not necessarily equal to the density outside, ρ', which is determined by the overall structure of the star. Whether the material continues to rise, or falls, is now determined by whether it is heavier or lighter than the material around ($\rho > \rho'$?).

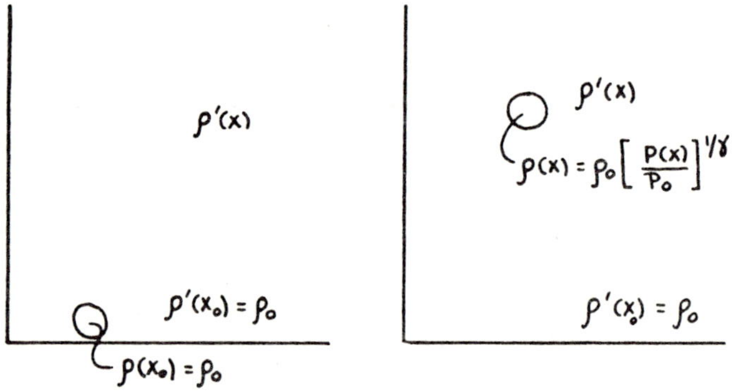

Fig. 3.2. Adiabatic Convection. An element of material at the density of the material around it rises adiabatically to the height x, where it has the density

$$\rho(x) = \rho_o \left[P(x)/P_o \right]^{1/\gamma} ,$$

determined by the adiabatic expansion law. The density $\rho'(x)$ of the material around the element is not necessarily equal to $\rho(x)$.

NEUTRINO ASTROPHYSICS

Now we write out the detailed equations. For the material rising

$$\rho/P(dP/d\rho)_{ad} = \gamma \tag{3.16}$$

or

$$dP/P = \gamma\, d\rho/\rho .$$

The material in the star is assumed to be an ideal gas, so

$$P = \rho' R_g T$$

and consequently the change in pressure in the star is related to the change in density and temperature in the star by

$$\delta P/P = \delta\rho'/\rho + \delta T/T . \tag{3.17}$$

Since $dP = \delta P$ (pressure inside = pressure outside)

$$\gamma\, d\rho/\rho = \delta\rho'/\rho' + \delta T/T .$$

If we have a mass of material at density $\rho_o(=\rho'_o)$ and move it up an infinitesimal amount, it will continue to rise if $d\rho < \delta\rho$ i.e., if the gas after it rises is lighter than the ambient gas. It will be stable if $d\rho > \delta\rho'$. Substituting in Eq. (3.17) above

$$\gamma\, \frac{\delta\rho'}{\rho'} \leq \frac{\delta\rho'}{\rho'} + \frac{\delta T}{T} ,$$

$$(\gamma-1)\frac{1}{\rho'}\frac{\delta\rho'}{\delta r} \leq \frac{1}{T}\frac{\delta T}{\delta r} , \tag{3.18}$$

or changing notation in an obvious way and noting that both the density and temperature decrease outward, the criterion for stability is that the temperature gradient cannot be too large, or precisely

$$\left|\frac{dT}{dr}\right| \leq (\gamma-1)\frac{T}{\rho}\left|\frac{d\rho}{dr}\right| . \tag{3.19}$$

This temperature gradient is called the adiabatic gradient.

Using values for the quantities appropriate to the earth's surface, one verifies that it is unstable against convection. This is the reason for the circulation of air and the transport of energy by winds.

The surface of the sun is unstable against convection, and there appear cells on the solar surface, offering evidence of the presence of convection. The cells are roughly 180 km in diameter, approximately the area that would be expected for a rising column of gas.

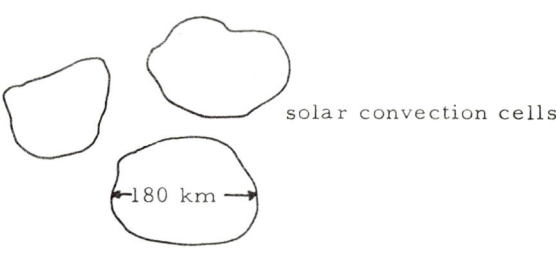

solar convection cells

One of the features of convection is that it is extremely efficient in transporting material. A star that is in convection equilibrium could bring matter that is created in the center to the surface in only a few hundred days. Observations of elements unstable against radioactive decay at the surface of stars is evidence both for nuclear genesis and convection. For the same reason the temperature gradient of stars will not be very much different from that given by Eq. (3.19).

Analytical Solutions

In general only trivial analytical solutions exist. The ones usually studied are called polytropes. The equation of state is:

$$P = \rho R_g T \quad (3.20)$$

and P is related to ρ and T by

$$P \propto \rho^{1+1/n}, \quad \rho \propto T^n, \quad P \propto T^{n+1} \quad (3.21)$$

If $n \to \infty$, then $P \propto \rho$, so we may say T = const. If n = 0, then ρ = const., $P \propto T$. n is called the "polytropic index."

Using the equation of hydrostatic equilibrium, together with the equations above, the "Lane-Emden equation of index n" may be derived:

$$\frac{1}{\xi^2}\frac{d}{d\xi}\left(\xi^2 \frac{d\Theta}{d\xi}\right) = -\Theta^n, \quad (3.22)$$

where ξ is proportional to r and $\rho_c \Theta^n = \rho$. This equation must be solved with the boundary conditions $\Theta = 1$ at $\xi = 0$. Analytical solutions of the equation exist for n = 0, n = 1 and n = 5.

For further reference, see Chapter IV, Chandrasekhar, Introduction to Stellar Structure.

4. NUCLEAR REACTIONS

4.1 Theory

The potential of a nucleus may be represented as follows:

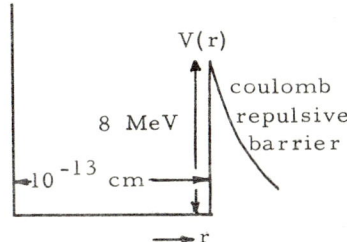

In order for a nuclear reaction to proceed, the two nuclei must come within the 10^{-13} cm range of the nuclear forces, penetrating through the Coulomb barrier. The kinetic energy available is $kT \sim 1 \text{KeV/particle} \ll 8$ MeV. Nevertheless, the transition probability is not zero. Consider the reaction $A + B \longrightarrow C + D$.

The transition probability β is the reciprocal of the mean reaction time τ and is given by

$$\beta = 1/\tau = \sigma v \qquad (4.1)$$

v is the relative velocity of two incident particles (three body reactions are not important in astrophysics), and σ, the cross section, is given by

$$\sigma = |\langle \psi_f | H_I | \psi_i \rangle|^2 \frac{2\pi}{\hbar} \rho(E_i) \qquad (4.2)$$

At low energy, if there are no resonant levels, the squared matrix element is a constant, apart from a Coulomb barrier factor ω which is given by

$$\omega = \frac{2\pi\eta}{\exp(2\pi\eta) - 1} \qquad (4.3)$$

$$\eta = \frac{1}{\hbar v}\left(2\pi Z_1 Z_2 e^2\right) = \frac{2\pi Z_1 Z_2}{137} \frac{c}{v} \qquad (4.4)$$

For hydrogen with energy ~ 1 KeV

$$\eta \sim 35 - 100 \gg 1$$

29

Therefore, we may write

$$\omega = 2\pi\eta\, e^{-2\pi\eta} \tag{4.5}$$

Where there is a resonant level,

$$\sigma = \frac{\pi \lambdabar^2 g\, \Gamma_{A+B}\, \Gamma_{C+D}}{(E_R - E)^2 + \Gamma_R^2/4} \; ; \tag{4.6}$$

g is a statistical factor, given by

$$g = \frac{2S_R + 1}{(2S_A + 1)(2S_B + 1)} \; . \tag{4.7}$$

Subscripts A, B, R refer to the spins (S) and energy (E) of the incident particles A and B and the resonant level R. This is the Breit-Wigner formula. Γ is the partial width.

Returning to the case of the non-resonant cross section, we may write it in the form

$$\sigma = \frac{S}{E} \exp\left[-2\pi \frac{Z_1 Z_2\, e^2}{\hbar v} \right] , \tag{4.8}$$

where E is the center of mass energy and S is a constant that measures the strength of the interaction due to the nuclear matrix elements. A typical value of S, for strong interactions, is the value for

$$B^{10}(p,\alpha)Be^8 \quad \text{i.e.} \quad B^{10} + p \rightarrow Be^8 + \alpha, \quad S \sim 10^7\, \text{eV barns.}$$

For a weak interaction, in particular $P(p, e^+\gamma)D$, $S \sim 10^{-19}$ eV barns.

The reaction rate dr is determined from the cross section by

$$dr = N_1(v_1)dv_1 N_2(v_2)dv_2\, \sigma(v)\, |\vec{v}_1 - \vec{v}_2| . \tag{4.9}$$

At the densities and pressures appropriate to a star at the early stages of nuclear evolution (when the proton-proton and carbon cycles are operating), the Maxwell distribution is appropriate for the number densities N_1 and N_2:

$$N_i = n_i \left(\frac{m_i}{2\pi kT}\right)^{3/2} \exp\left(-\frac{\frac{1}{2}m_i v_i^2}{kT}\right) dv_i . \quad (4.10)$$

The reaction rate is

$$r = n_1 n_2 \left[\frac{\sqrt{m_1 m_2}}{2\pi kT}\right]^3 \int \exp\left\{-\frac{1}{kT}\left[\sum \frac{m_i v_i^2}{2}\right]\right\} \sigma(v) |v| dv_1 dv_2 . \quad (4.11)$$

We change variables to relative velocity and c.m. velocity

$$\vec{v} = \vec{v}_1 - \vec{v}_2 ,$$
$$\vec{V} = \frac{m_1 \vec{v}_1 + m_2 \vec{v}_2}{m_1 + m_2} \quad (4.12)$$

The Jacobian of the transformation is easily verified to be unity:

$$dv_1 dv_2 = \frac{\partial(v_1 v_2)}{\partial(v V)} dv dV = dv dV \quad (4.13)$$

Then

$$r = n_1 n_2 \left(\frac{\sqrt{m_1 m_2}}{2\pi kT}\right)^3 \int \exp\left\{-\frac{1}{kT}\left[\left(\sum_i m_i\right)\frac{V^2}{2} + \frac{m_1 m_2}{m_1 + m_2} v^2\right]\right\} \quad (4.14)$$

$$\times \sigma(v) v \, dv \, dV .$$

The V integration can be done, giving

$$r = 4\pi n_1 n_2 \left(\frac{m_1 m_2}{2\pi kT(m_1 + m_2)}\right)^{3/2} \int_0^\infty dv \, v^3 \sigma(v) \, e^{-\frac{m_1 m_2}{(m_1 + m_2)}\frac{v^2}{kT}} . \quad (4.15)$$

If we substitute now the previous expression for the cross section, and change the independent variable from v to the cm energy $E \propto v^2$, then

$$r = 2 n_1 n_2 \left(\frac{2}{\pi \mu}\right)^{1/2} \frac{S}{(kT)^{3/2}} \int_0^\infty dE \, \exp\left\{-\frac{2E}{kT} - \frac{B}{E^{1/2}}\right\} . \quad (4.16)$$

where

$$B = \frac{2\pi Z_1 Z_2 e^2}{\hbar} \left(\frac{\mu}{2}\right)^{1/2}$$

$$\mu = \frac{m_1 m_2}{m_1 + m_2} = \text{the reduced mass.}$$

To evaluate this integral, we first change to dimensionless variables. In this way the dependence on the physical quantities is made explicit, and the integral becomes a number. Then we approximate the integral by the method of steepest descent. Let $x = 2E/kT$; then the integral to be done is

$$\int_0^\infty \exp - [x + a/\sqrt{x}] \, dx \, . \qquad (4.17)$$

The maximum occurs at $1 - \frac{1}{2} a/x^{3/2} = 0$.
Let $\xi = x - (a/2)^{2/3}$, and the integral is

$$2\left(\frac{1}{3}\pi\right)^{1/2} \left(\frac{1}{2}a\right)^{1/3} \exp\left(-3\left(\frac{1}{2}a\right)^{2/3}\right); \qquad (4.18)$$

$$a = 1.066 \times 10^2 Z_1 Z_2 \left(\frac{A_1 A_2}{A_1 + A_2}\right)^{1/2} \frac{1}{T_6^{1/2}} \, ,$$

$$T_6 = T/10^6 \, .$$

The most convenient way to express the final result is in terms of the percent by weight of species one, X, and the reaction rate per nucleus of type two, P:

$$n_1 = \frac{N_0 \rho}{A_1 m_p} X_1 \, , \qquad (N_0 = \text{Avogadro's number})$$

$$P = r/n_2 = 434 \rho X_1 \left(\frac{A_1 + A_2}{A_1^2 A_2^2 Z_1 Z_2}\right) S\tau^2 e^{-\tau} \, , \qquad (4.19)$$

NEUTRINO ASTROPHYSICS 33

$$\tau = 42.5 \left[\frac{Z_1^2 Z_2^2 A_1 A_2}{A_1 + A_2} \right]^{1/3} T^{-1/3} . \qquad (4.20)$$

The temperature dependence of the rate does not depend upon S. The maximum reaction rate occurs at a temperature determined independently of S. The macroscopic structure of stars is essentially independent of the details of nuclear collisions, and the value of the cross section is important primarily in determining the details of nuclear synthesis.

We can write

$$r = A \rho X_H T^{-2/3} \exp{-B/T^{1/3}} \qquad (4.21)$$

and summarize the rates of various reactions in the table below.

	Reaction	S	A	B
1.	$H^1 + H^1 \rightarrow D^2 + e^+ + \nu$	3.4×10^{-22}	3.3×10^{-13}	33.8
2.	$D + H^1 \rightarrow He^3 + \gamma$	3×10^{-4}	2.4×10^5	37.2
3.	$Li^7 + P \rightarrow 2 He^4$	1.0×10^2	1.2×10^{11}	84.7
4.	$C^{12} + P \rightarrow N^{13} + \gamma$	1.2	1.7×10^9	136.9
5.	$N^{14} + P \rightarrow O^{15} + \gamma$	3.1	4.7×10^9	152.3

Source: Hubert Reeves, <u>Stellar Energy Sources.</u>

The units of S are KeV-barns. The error is in the last significant figure in S and A. B is accurate to better than 0.001%.

Note the difference in the value of S for the weak interaction (1), the electromagnetic interactions (2, 4, 5) and the strong interaction (3).

4.2 Specific Reactions

(a) The Proton-Proton Reactions

The fundamental nuclear reaction is the proton-proton reaction

$$P + P \longrightarrow D^2 + e^+ + \nu.$$

For low energy collisions the two protons are in an S state and the spin wave function is therefore odd (singlet state). In the final state (deuteron), the isotopic spin wave function is odd, the spatial wave function contains S and D waves, and is even, and the spin wave function is even. Both states have even parity. These quantum numbers determine the transition to be of the allowed Gamow-Teller type.

The transition probability β is

$$\beta = |\langle \Psi_f | H_I | \Psi_i \rangle|^2 \frac{2\pi}{\hbar} \rho(\epsilon) ,$$

$\rho(E)$ = density of final states.

Beta decay theory gives

$$\beta = 4 \frac{G^2}{2\pi^3} \frac{m_e c^2}{\hbar^2} f(w) \left| \int \Psi_i \Psi_f \, d\tau \right|^2 ,$$

$$f(w) = (w^2-1)^{1/2} \left(\frac{1}{30} w^4 - \frac{3}{20} w^2 - \frac{2}{15} \right) \qquad (4.22)$$
$$+ \frac{w}{4} \ln \left[w + (w^2-1)^{1/2} \right].$$

W is the maximum energy of the electron in units of $m_e c^2$, and the integral is over the spatial wave functions. G is the β-decay coupling constant.

(b) Light Element Synthesis

The chain of reaction in the stars will begin in the following way:

NEUTRINO ASTROPHYSICS

$P + P \longrightarrow D + e^+ + \nu$ Weak interaction; goes slowly.

$D + P \longrightarrow He^3 + \gamma$ Only 2.3 sec.

$He^3 + P \longrightarrow He^4 + e^+ + \nu$ This is a weak interaction and goes slowly. Consequently the He^3 builds up until it can go by:

$He^3 + He^3 \longrightarrow He^4 + 2P$

$He^3 + He^4 \longrightarrow Be^7$

$Be^7 + e^- \longrightarrow Li^7 + \nu$

$Li^7 + P \longrightarrow Be^8 \longrightarrow 2 He^4$ Be^8 does not have a chance to build up before it decays.

Is there any way, by this series of reactions, to build up the heavy elements?

$Li^6 + P \longrightarrow He^4 + He^3$

$Li^7 + D \longrightarrow 2 He^4 + n$

$Be^9 + D \longrightarrow Li^7 + He^4$

$ \longrightarrow 2 He^4 + He^3$

$ \longrightarrow Be^{10} + P$

$Be^{10} + P \longrightarrow Be^{11}$

$Be^{11} + P \longrightarrow 3 He^4$

Thus, none of these reactions could succeed in building heavy elements. Also worth noting is that the last group of reactions proceed at fairly low temperature and would burn out Be and Li before the star ever reaches the proton burning stage, if these elements existed in the proto-star. The observation of these elements on the surface of stars, then, must be evidence for their continuing synthesis, either inside the star, if there is mixing, or if there is no mixing, on the surface by means of protons accelerated from the center.

(c) Carbon Cycle

We mentioned in the last lecture the reactions which follow the proton-proton reaction. The result of these reactions is to synthesize He^4 from hydrogen, with a resulting liberation of energy.

If we assume that C^{12} exists in stellar materials, then there is another group of reactions, known as the carbon cycle, which occur at stellar temperatures, which synthesize He^4 from protons using carbon as a catalyst.

$C^{12} + P \longrightarrow N^{13} + \gamma$

$N^{13} \longrightarrow C^{13} + e^+ + \nu$ Slow, ~100 sec, a weak interaction

$C^{13} + P \longrightarrow N^{14} + \gamma$

$N^{14} + P \longrightarrow O^{15} + \gamma$

$O^{15} \longrightarrow N^{15} + e^+ + \nu$

$N^{15} + P \longrightarrow O^{16} + \gamma$ The ratio of these two branches is

$\phantom{N^{15} + P} \longrightarrow C^{12} + He^4$ $1:10^4$

$N^{14} + P \longrightarrow O^{15} + \gamma$

$O^{15} \longrightarrow N^{15} + e^+ + \nu$

$N^{15} + P \longrightarrow O^{16} + \gamma$

$O^{16} + P \longrightarrow F^{17} + \gamma$

$F^{17} \longrightarrow O^{17} + e^+ + \nu$

$O^{17} + P \longrightarrow N^{14} + He^4$

These reactions go at a slightly higher temperature so that some of the carbon is taken out of the cycle. Each carbon catalyses about $100\ H - 25\ He^4$ in the lifetime of the star. About 1% of the carbon present becomes O^{16}.

(d) Synthesis of Carbon

We have seen that there are no proton reactions which can synthesize carbon. Let us consider He^4 reactions. As a result of proton burning, a considerable concentration of He^4 can be built up.

The three body reaction

$$He^4 + He^4 + He^4 \longrightarrow C^{12}$$

is too slow to be the source of carbon. But Be^8, while unstable, has a longer lifetime than the time between He^4 collisions. Nevertheless, the two-step reaction

$$He^4 + He^4 \rightleftharpoons Be^8$$
$$Be^8 + He^4 \longrightarrow C^{12}$$

is also too slow to be the source of stellar carbon.

Believing that carbon had to be made too within stars, Hoyle postulated the existence of a resonant excited state of carbon. Later a resonant level was discovered at 7.64 MeV, very close to the Be^8-He^4 system energy.

The actual process is the three-step reaction

1. $He^4 + He^4 \rightleftharpoons Be^8$ — 94 KeV
2. $Be^8 + He^4 \rightleftharpoons C_{12}^*$ — 278 KeV
3. $C^{12*} \longrightarrow C^{12} + 2\gamma$ + 7.654 MeV

The number density $n_{C_{12}^*}$ of C_{12}^* is given by the law of mass action as

$$n_{C_{12}^*} = n_\alpha^3 \left(\frac{2\pi \hbar^2}{M_\alpha \, kT} \sqrt{3}\right)^3 e^{-[94 + 278]/kT} \quad ;$$

n_α and M_α are the number density and mass of the alpha particle, kT is expressed in units of KeV.

The number, P, of C^{12} nuclei formed per second is, then,

$$P = n_{C_{12}^*} / \tau \quad ,$$

where τ is the mean decay time for reaction 3 above. $\tau \sim 10^{-11}$ sec.

(e) α-Process; Carbon, Oxygen Burning

Once carbon has been produced, a number of higher elements can be produced by helium bombardment, the "α-process."

$$C^{12} + He^4 \longrightarrow O^{16}$$

$$O^{16} + He^4 \longrightarrow Ne^{20}$$

$$Ne^{20} + He^4 \longrightarrow Mg^{24}$$

Further bombardment can produce similar nuclei all the way to $_{22}Ti^{44}$ and $_{22}Ti^{48}$.

Another process that can be important is

$$C^{12} + C^{12} \longrightarrow Mg^{24}$$
$$\longrightarrow Na^{23} + P.$$

Also

$$O^{16} + O^{16} \longrightarrow S^{32*}.$$

The photon energy inside stars is about 0.05 MeV at temperatures appropriate to the evolutionary stage toward the end of the carbon cycle. At these energies some of the higher elements begin to undergo photo-disintegration, and the situation becomes messy and chaotic. At higher temperatures the abundances of elements are determined by the statistical equilibrium.

4.3 Pycnonuclear Reactions

At zero temperature matter must be in a crystalline form. This may be seen from the third law of thermodynamics, which requires that at $T = 0$ matter be ordered.

At zero temperature, moreover, the free energy must be a minimum. Consequently the Coulomb energy must also be a minimum. Let $r = \bar{r} + (r - \bar{r}) = \bar{r} + \Delta r$. The Coulomb energy is

$$\frac{Ze^2}{r} = \frac{Ze^2}{\bar{r}} \frac{1}{1 + \frac{\Delta r}{\bar{r}}} \approx \frac{Ze^2}{\bar{r}} \left[1 - \frac{r - \bar{r}}{\bar{r}} + \left(\frac{r - \bar{r}}{\bar{r}}\right)^2 \right]. \quad (4.23)$$

If we average this over the distance between one atom and each of the others, clearly $\overline{\frac{r - \bar{r}}{\bar{r}}} = 0$. Therefore

NEUTRINO ASTROPHYSICS

$$\frac{1}{N}\sum_{j}^{N}\frac{Ze^2}{r_{ij}} = \frac{Ze^2}{\bar{r}}\left(1 + \frac{\overline{(\Delta r)^2}}{\bar{r}^2}\right) . \qquad (4.24)$$

So, to obtain the minimum Coulomb potential, one must find the crystal structure with the smallest mean square deviation. The face centered lattice has the smallest Coulomb energy in this sense. One must now compute the "Madelung sum,"

$$\sum_{i \ne j} \varphi_{ij} \quad ; \quad \varphi_{ij} = \frac{Z^2 e^2}{r_{ij}} .$$

For the face centered lattice, the result turns out to be

$$\sum_{i \ne j} \varphi_{ij} = 1.8 \frac{Z^{2/3}}{r_e}\left(\frac{\alpha^2 mc^2}{2}\right) \qquad (4.25)$$

$$\frac{\alpha^2 mc^2}{2} = 1 \text{ rydberg} = 13.6 \text{ eV},$$

r_e is the interelectronic distance.

The configuration is a lattice of positive ions, with each ion surrounded by a lattice of electrons.

At high density this Coulomb energy will cause zero point motion, leading to reactions even without thermal motion. Moreover, electron concentration will raise the Coulomb energy level of the protons. Hence even at zero temperature the ions will have some energy due to the electrical properties of the medium. Nuclear reactions thus can proceed at zero temperature and high density. Such reactions are called pycnonuclear or pressure induced nuclear reactions.

To compute the electric field around the ion, the lattice of negative charge about each ion is replaced by a sphere of uniformly distributed charge. Because of the high crystallographic symmetry, this approximation, introduced by Wigner and Seitz, is very good. (The Coulomb energy computed by the Wigner-Seitz sphere method comes to within 0.2% of the value evaluated according to the Madelung sum method for a face centered cubic crystal lattice which has a minimum in Coulomb energy.)

If the ion moves about in its cell a distance r from the center, it feels a potential due to the electrons within r,

$$V(r) = \frac{e^2}{r}\rho \cdot \frac{4}{3}\pi r^3 = \frac{4}{3}\pi \rho e^2 r^2 \quad ; \qquad (4.26)$$

ρ is the density of electrons. This is a simple harmonic motion potential. The ground state energy is

$$E_o = 3/2 \hbar \omega \qquad (4.27)$$

where ω is the angular frequency. The energy per electron is

$$E_{zero\ point} = 3 \left(\frac{m_e}{Z\ M_{ion}}\right)^{1/2} \frac{1}{r_e^{3/2}} ; \qquad (4.28)$$

r_e is the inter-electronic distance in units of $\hbar^2/me^2$.

The potential of one ion with respect to another entering the Wigner-Seitz sphere from infinity is

$$V = \frac{Ze}{r} - \frac{Ze}{R} + \frac{Ze}{R}\left(\frac{r}{R}\right)^2, \qquad r < R ,$$

where R is the radius of the sphere. Hence, at $r \approx 0$, the electrostatic repulsive potential between two ions will be decreased by an amount Ze/R which depends on the density only. This term Ze/R can be lumped with the energy term in the Schrödinger equation and the ion seems to gain an extra energy of Ze/R upon entering the Wigner-Seitz sphere.

5. THE HERTZSPRUNG-RUSSELL DIAGRAM

These equations determine the structure of a star:

$$\frac{dP}{dr} = -\frac{Gm}{r}\rho \tag{5.1}$$

$$\frac{dm}{dr} = 4\pi r^2 \rho \tag{5.2}$$

$$\frac{dL}{dr} = 4\pi r^2 \rho \epsilon \tag{5.3}$$

The equilibrium equations for a body in radiative equilibrium.

$$T^3 \frac{dT}{dr} = \frac{4}{3}\frac{ac}{K\rho}\frac{L}{4\pi r^2} \tag{5.4}$$

$P = P(\rho, T)$ The equation of state. (5.5)

$\epsilon = \epsilon(\rho, T)$ The energy generation, determined from nuclear physics. (5.6)

$K = K(\rho, G)$ The opacity, given by atomic theory and thermodynamics. (5.7)

With these equations there is a unique relation between R, L, M, and T_S. The mass is put in, as it depends on the conditions of formation of the proto-star, and there is a relation among the remaining variables,

$$L = 4\pi R^2 ac T_S^4 \;; \qquad T_S \text{ is the surface temperature.} \tag{5.8}$$

One can plot the observed luminosity against the surface temperature of stars. Such a diagram, called the Hertzsprung-Russell diagram, is shown in Fig. 5.1.

The mass is increasing toward the higher values of luminosity.
One might ask what path is traced out by an evolving star in its lifetime. This path is not along the main sequence as might be thought. We shall return to this.

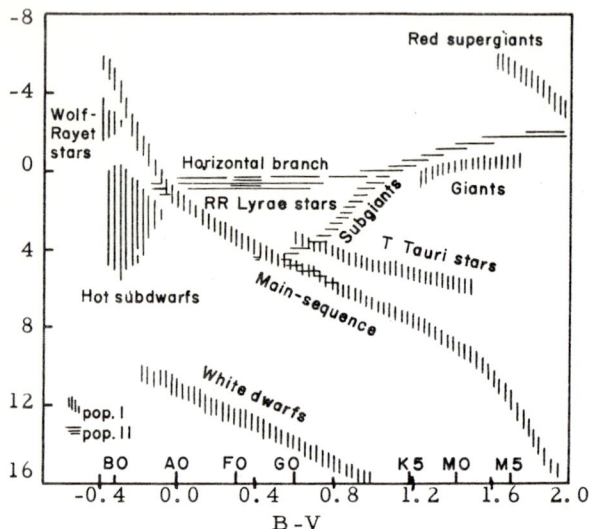

Fig. 5.1. Schematic HR diagram of stars, showing the locations of various sequences. The spectral types are for the main-sequence stars. [Source: Hayashi, Hoshi, and Sugimoto, Evolution of the Stars, p. 8.]

For stars along the main sequence $R \propto R_\odot \, M/M_\odot$. The functional relation between the mass and the luminosity is

$L \propto M^3$,   middle of main sequence ,   (5.9)

$L \propto M^5$ along the ends of the main sequence .

For the red giants the radii are $\sim 10^{13}$ cm. The cores of these stars are as small as the earth, but the surface extends to the distance of the earth's orbit. Aside from the indirect measurement of the radii by measuring the surface temperature and absolute luminosity, applying Eq. (5.8) above, a direct measure of the radius can be obtained by measuring the size of interference fringes produced from the light of some of these giant stars.*

NEUTRINO ASTROPHYSICS

It is interesting to note that the density of red giants cannot be ~ 1 gm/cc, for if this were the case, then the red shift is of the order of

$$\frac{GM^2}{R} \cdot \frac{1}{Mc^2} = \frac{GM}{Rc^2} = \frac{G\rho R^2}{c^2} \sim \frac{10^{-7}(10^{13})^2}{10^{21}} \sim 10^{-2} \quad . \tag{5.10}$$

If gravitational energy were such a substantial fraction of the rest energy, then the gravitational red shift would be larger than observed. The observed red shift corresponds to a density of about 10^{-3} g/cc. Actually $M \sim M_\odot$.

We will not discuss variable stars, except to say that they occur in several types, having periods of a few hours, days, or longer. The oscillations of the radii, interestingly, seem to be almost opposite in phase to the oscillations of the luminosity. Furthermore, the period of variables is related to their absolute luminosity. For Cepheid variables, stars having periods of the order of days, this is approximately

$$L \propto T^{1/2} \tag{5.11}$$

with the exponent of T actually increasing slightly with increasing L. Because of this relation, it is possible to measure the distances of stars by finding a variable star in the same group, measuring the period of the variable, using this to determine the absolute luminosity, and obtaining the distance by a comparison of absolute to observed luminosity. In this way the distance to Andromeda was measured as 1.7×10^6 light years.

White dwarfs are characterized by a mass $M \sim M_\odot$ and a radius $R \sim 10^9$ cm, approximately the earth radius. The first white dwarf was discovered by noticing motion of Sirius with respect to the background of stars, with a period of about 50 years. The sensitivity of this method is considerable.

One second of arc corresponds to $1/57 \times 1/60 \times 1/60 \sim (1/2 \times 10^5)$ radians. Motion ~ 0.01 seconds of arc is teloscopically observable. A perturbation of one solar radius $\sim 10^{11}$ cm is observable at a distance, D, of $0.01 \times \dfrac{1}{2 \times 10^5} = \dfrac{10^{11}}{D}$. $D = 10^{18}$ cm = 1 light year. Thus, the existence of Jupiter could be surmised by observing the sun from a distance of one light year.

The white dwarf companion to Sirius was first seen by Alven Clark. It was not realized at first how important this star was, until measurements of temperature and luminosity led to determination of the radius, and, with the value of its mass known from oribtal data, the density was found to be 10^5 g/cc. (The density of lead is 11 g/cc.)

A theory of the interior of these stars was worked out by Eddington, then substantially modified by Chandrasekhar, in which the weight of the material is supported by the Fermi pressure, and the star is a degenerate Fermi gas. This is true for all except a very thin surface atmosphere. We compare the atmosphere of the white dwarf with that of other stars.

First, we note that in any case the atmosphere will be small compared to the total radius, so that we may approximate the atmosphere as planar and use the density formula

$$\rho = \rho_0 \, e^{-z/\left(\frac{kT}{mg}\right)} . \qquad (5.12)$$

$z_0 = \frac{kT}{mg}$ is called the scale height and, being the distance over which the density drops by a factor of e, is essentially the height of the atmosphere.

At the surface of the sun, $g = 27$, $T \sim 10^4$; therefore

$$z_0 = \frac{kT}{mg} = \frac{R_g T}{g} \sim \frac{10^8 \cdot 10^4}{10^4} \sim 10^8 \text{ cm} = 1000 \text{ km}. \qquad (5.13)$$

By contrast, for a white dwarf, since $g = GM/R^2$ and the mass is about solar mass while the radius is much smaller, $g \sim 10^9$, then $z_0 \sim 10^3$ cm. For a red giant the atmosphere may range from 10^9 to 10^{13} cm.

The evolution of a star in the Hertzsprung-Russell diagram, as mentioned before, is not along the main sequence. Instead, the star starts to the right of the main sequence as a protostar. The temperature increases during the phase of gravitational contraction and the star becomes somewhat brighter for awhile before it reaches the main sequence. It then remains at the same point for most of its life, undergoing the proton-proton reactions and the carbon cycle. The center fills up with a degenerate core of He^4, with the hydrogen burning proceeding at the boundary of the core until the condition is reached

$$\frac{M_{He^4}}{M_{star}} \sim 0.6 .$$

The star, at about this time, begins to expand and new nuclear processes begin to occur, as the temperature climbs about an order of magnitude to $\sim 10^8$ °K, until the star becomes a red giant. A typical path to this stage of evolution is shown in Fig. 5.2. Dots indicate where the length of time spent is much longer than places where there are arrows.

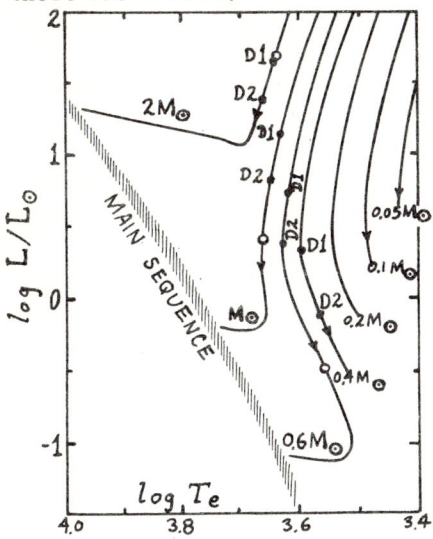

Fig. 5.2. Evolutionary Tracks of Contracting Population I Stars. The open circles denote the ends of the wholly convective stages. The stages of deuterium burning are indicated by D1 and D2.

An estimate of the lifetime T of a star can be made by using the relation

$$T = \frac{\text{Available energy}}{\text{Rate of energy emission}}.$$

The available energy is ~ 0.01 Mc^2 because the binding energy of iron is about 8 MeV/nucleon, and it is the most stable nuclei. Hence, for each 900 MeV of the mass of the star, at most about 8 MeV can be liberated by nuclear reactions. The luminosity L is exactly the rate of energy emission, hence

$$T \sim \frac{0.001 \text{ Mc}^2}{L}.$$

This relation gives a lifetime $\sim 10^{10}$ years for the sun. Since $L \propto M^{3-5}$, heavier stars are brighter stars and burn out faster. A star with $L \sim 10^3 L_\odot$ would live for $T \sim 10^7$ years.

Stars exist of a wide range of ages. The Pleiades are about 10^5 years old, while stars in a cluster called NGC 168 are $(20 \pm 4) \times 10^9$ years old, with the limits due to uncertainty in the knowledge of the concentration of metals in these stars.

Lifetimes greater than 10^{10} years are suspected, because of a lack of corroborating evidence for such large ages. For other reasons it has been suggested by Dirac that the value of G, the gravitational constant, is changing with time. Dicke has applied this to the evolution of stars and concludes that with such an effect the age of stars in NGC 168 becomes $\sim 8 \times 10^9$ years.

Up to this stage of evolution the physics has depended primarily upon classical gravitation and equilibrium theory for the gross static structure, nuclear and atomic physics for complete determination of the equations, the factors ϵ and K in the radiative equilibrium equations. Stellar evolution is shown pictorially in Fig. 5.3.

For the later stages of evolution neutrino physics and general relativity become important. So we now leave the discussion of evolution for a discussion of these topics.

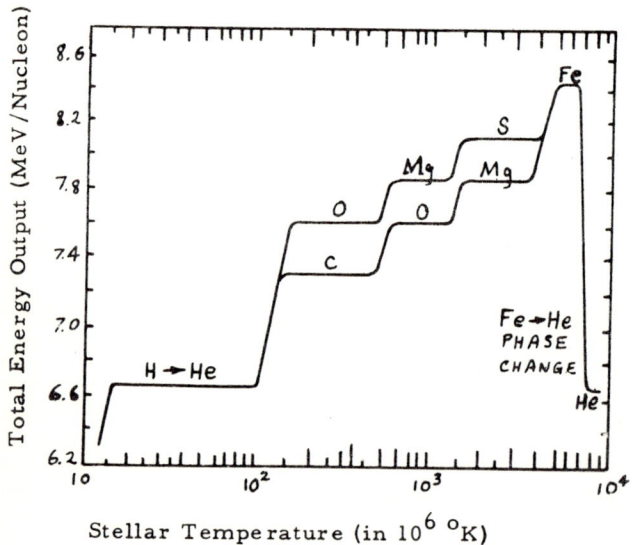

Fig. 5.3. Nucleosynthesis in Stellar Evolution.

6. NEUTRINO PROCESSES

The most important neutrino processes are:
1. The URCA process;
2. Neutrino bremsstrahlung;
3. The photo-neutrino process;
4. Pair annihilation to neutrinos;
5. Neutrino emission from a plasma.

The reason neutrino processes are important is that once neutrinos are produced, they escape from the star, acting as an energy sink. The mean free path λ for neutrinos is given as a function of the cross section σ (which is of the order of 10^{-44} cm^2 for processes which could stop the neutrino) and the matter density n (about 10^{23} particles/cc) by

$$\lambda = 1/\sigma n \sim 1/10^{-44} \cdot 10^{23} \sim 10^{21} \text{ cm} \sim 1000 \text{ light years.} \quad (6.1)$$

Since the radius of stars is less than $\sim 10^{13}$ cm, the star is completely transparent to the neutrinos produced inside. By contrast, the photons produced diffuse slowly from the star, contributing meanwhile to the total pressure.

The neutrino luminosity L is

$$L_\nu = \int \epsilon_\nu \, dV, \quad (6.2)$$

where ϵ_ν is the neutrino energy production rate per cc.

6.1 Neutrino Processes

(a) The URCA Process

The loss of energy due to ordinary β decay is called the URCA process

$$(Z, A) + e^- \rightleftharpoons (Z-1, A) + \nu .$$

If the equilibrium is dynamic, then energy can be constantly drained away by this process. To see the order of magnitude of the importance of this process, we consider two examples. Mn^{56} decays to Fe^{56} in 2.5 hours. The neutrinos carry a mean energy of 1 MeV. In an Mn^{56} star the rate of energy loss to neutrinos is

$$\frac{\dfrac{1\text{ MeV}}{\text{nuclei}} \times \dfrac{1}{56}\dfrac{\text{moles}}{\text{gm}} \times 10^{23}\dfrac{\text{nuclei}}{\text{mole}} \times 10^{-6}\dfrac{\text{erg}}{\text{MeV}}}{2.5 \times 3600 \text{ sec}} = \frac{2 \times 10^{11} \text{ ergs}}{\text{gm-sec}}. \quad (6.3)$$

N^{13} decays to C^{13}. For a dynamic process to be established, the inverse β-decay would have to occur,

$$C^{13} + e^- \longrightarrow N^{13} + \nu ,$$

but this reaction requires 8 MeV, while in a star, the energies available are thermal and only ~ 0.5 MeV. This indicates that the URCA process will be ineffectual due to a lack of dynamic equilibrium.

Nuclei of charge $Z-1$ decay at the rate

$$\left(\frac{dn_{Z-1}}{dt}\right)_{decay} = \lambda_{Z-1} n_{Z-1}. \quad (6.4)$$

They are produced at the rate

$$\left(\frac{dn_{Z-1}}{dt}\right)_{production} = \int N_Z(p_Z) N_e(p_e) \sigma v \, d^3 p_Z \, d^3 p_e . \quad (6.5)$$

The cross section is given in terms of the observed ft value by

$$\sigma = \frac{2\pi^2 \log 2}{ft} \frac{\hbar^3}{m^3 c^4} \frac{(E - E_{max})^2/(mc^2)^2}{v/c} , \quad (6.6)$$

where f is a calculable function of the energy and t is the half-life (see Eq. (4-22)). The product ft is related to the coupling constant and matrix element by

$$g^2 |M|^2 = 2\pi^3 \ln 2/ft . \quad (6.7)$$

E is the energy of the electron. E_{max} is the maximum energy with which the electron could emerge in ordinary β-decay, consequently here the minimum energy which permits a neutrino and allows conservation of energy and momentum. $(E - E_{max})$ is the neutrino energy.

NEUTRINO ASTROPHYSICS 49

The electron density is given by the Boltzmann distribution:

$$N_e = n_e \left(\frac{m_e}{2\pi kT}\right)^{3/2} \exp(-\frac{1}{2} mv^2/kT) \ . \tag{6.8}$$

The nuclei, because they are heavy, can be taken as stationary. Putting this in,

$$\left(\frac{dn_{Z-1}}{dt}\right)_{production} = 4\pi n_Z \int_E^\infty N_e \sigma v\, p_e^2 dp_e \tag{6.9}$$

$$= 4\pi n_e n_Z \left(\frac{m_e}{4\pi kT}\right)^{3/2} e^{-E_{max}/kT} \int_0^\infty v\sigma(E_\nu)(E_{max} + E_\nu)^2$$

$$\times e^{-E_\nu/kT} dE_\nu \ .$$

At $T = 10^9 \ °K$, $kT = 0.1$ MeV, $E_{max} \sim 7$ MeV. Thus, the exponential outside damps the production rate by a factor e^{-70}!

Most elements, as we just indicated, are either stable with a threhsold against inverse β-decay that is high, or unstable and decay rapidly to something else. There are few elements with a dynamic equilibrium between the decay and production products.

The elements which contribute most to the URCA process are Cl^{35} and S^{32}.

(b) <u>Neutrino Bremsstrahlung</u>

The theory of weak interactions of Feynman and Gell-Mann, as well as later ideas about weak interactions, such as the intermediate vector boson, allow not only for the reactions

$$n \rightarrow p + e^- + \bar{\nu}_e$$

and
$$\mu^- \rightarrow e^- + \bar{\nu}_e + \nu_\mu$$

but also for other reactions, which have not been observed yet in the laboratory.

The theory of Feynman and Gell-Mann describes the weak interactions as produced by a current which interacts with itself. The current

$$J = (e\,\nu_e) + (pn) + (\mu\,\nu_\mu)$$

interacting with itself produces, as well as the two reactions above, the reactions

$$(p, n)\,(p, n)^+$$
$$(\mu,\nu_\mu)\,(\mu,\nu_\mu)^+$$
$$(p, n)\,(\mu,\nu_\mu)^+$$
$$(e,\nu_e)\,(e,\nu_e)^+$$

Because the mass of the μ is much greater than that of the electron, muon processes are much less important than electron processes, contributing nothing substantially more important than the URCA process.

The square term processes $(\mu\,\nu_\mu)(\mu\,\nu_\mu)^+$ and $(e\,\nu_e)(e\,\nu_e)^+$ become important when the thermal energy kT is of the order of the μ mass and electron mass, respectively, because then pairs are in equilibrium with the photon field.

For μ, $\quad m \sim 106$ MeV, $\quad T \sim 10^{12}\,°K$.

For e, $\quad m \sim 0.5$ MeV, $\quad T \sim 7 \times 10^{9}\,°K$.

The first process that can be considered is neutrino bremsstrahlung with this diagram:

From the weak interaction there is a contribution to the cross-section $g^2 E_\nu^2$ and from the two photon vertices a factor of α^2. The cross section can be estimated as

$$\sigma \sim (g^2 E_\nu^2)\,\alpha^2 \sim 10^{-44}\,10^{-4} \sim 10^{-48}\,cm^2.$$

NEUTRINO ASTROPHYSICS

The rate of the reaction is calculated in the same manner as for nuclear reactions.

$$r = n_e n(Z,A) \sigma v , \qquad (6.10)$$

$$n_e \sim n(Z,A) \sim 10^{23}/cc ,$$

$$v \sim c \sim 3 \times 10^{10} \text{ cm/sec}.$$

The reaction rate is

$$r = n_e n(Z,A) \sigma v \sim 10^{23} \cdot 10^{23} \cdot 10^{-48} \cdot 10^{10} = \frac{10^3}{\text{sec cc}} . \qquad (6.11)$$

The energy of these neutrinos is about 0.1 MeV. So the energy production can be estimated as

$$E = 10^8 \cdot (0.1) \times 10^{-6} \sim 10 \text{ erg/cc-sec}.$$

It turns out, however, when the reaction is calculated in detail, that there is a numerical factor of $1/(35\,\pi^3) \sim 10^{-5}$, which drops the rate to where it is not important. This illustrates the fallacy of relying too heavily on order of magnitude estimates.

(c) Photo-Neutrino Process

We can find a reaction with a higher rate by eliminating one power of $\alpha = 1/137$, by elimination of one photon vertex, if we consider a real instead of virtual photon,

$$\gamma + e^- \longrightarrow e^- + \bar{\nu}_e + \nu_e .$$

This is the photoneutrino process.

There are two diagrams by which this interaction can proceed:

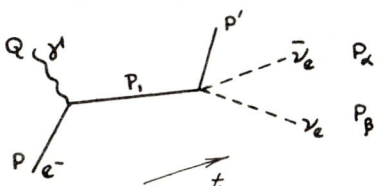

and

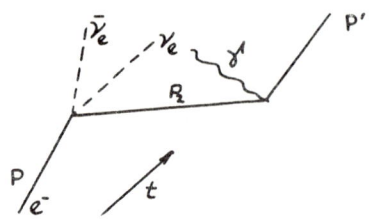

The matrix element is

$$M = M_1 + M_2 = -\sqrt{8}\, Ge(\bar{u}_\alpha \gamma_\mu . a u_\beta) \tag{6.12}$$

$$\left\{ (\bar{u}_{p'} \gamma_\mu a \frac{1}{\not{p}_1 - m} \not{q}\, u_p) + (\bar{u}_{p'} \not{q} \frac{1}{\not{p}_2 - m} \gamma_\mu a u_p) \right\},$$

$$a = \frac{1}{2}(1 + i\gamma_5).$$

With the aid of Fierz transformations, and with the spinors normalized so that $u^+ u = 2E$, the matrix element squared and summed over final spin states and averaged over initial spin states can be calculated straightforwardly, and is

$$|M|^2 = |M_1 + M_2|^2 \tag{6.13}$$

$$= \frac{1}{2}(8G^2)e^2 \text{Tr}[\gamma_\mu a \not{p}_\alpha \gamma_\nu a \not{p}_\beta] \times \text{Tr}[\overline{\mathcal{m}}(\not{p}'+m)\mathcal{m}(\not{p}+m)];$$

$$\mathcal{m} = \frac{\gamma_\mu a(\not{p}_1 + m)}{p_1^2 - m^2} + \frac{(\not{p}_2 + m)\gamma_\mu a}{p_2^2 - m^2}, \tag{6.14}$$

$$p_1 = P + Q, \quad p_2 = P' - Q.$$

To compute the cross section, the matrix element must be integrated over the available final states. For neutrinos, all states are available, but for the electrons, by the Pauli principle, since the electron density is n_e, the only available states are given by $1 - n_e(P)$. Thus the cross section is

NEUTRINO ASTROPHYSICS

$$\sigma = \frac{2\pi}{2E\,2q}\int \frac{1}{2}\sum_{pol}|M|^2 (2\pi)^3 \delta^4(P + Q - (P' + P_\alpha + P_\beta))$$

$$\frac{1}{(2\pi)^4}\frac{d^3P}{2E_\alpha}\frac{d^3P_\beta}{2E_\beta}(1 - n_e(P'))\frac{d^3P'}{2E'} \quad . \tag{6.15}$$

Since, in general, the outgoing electron has less energy than the incoming electron, the effects of the exclusion principle decrease the neutrino production rate.

The energy loss rate dE/dt is given by

$$-dE/dt = c\int N_\gamma(P_\gamma)N_e(P_e)E_\nu d\sigma\, d^3P_e d^3P_\gamma ; \tag{6.16}$$

$$E_\nu = E_{\gamma_\alpha} + E_{\gamma_\beta} = E + E_\gamma - E'.$$

Therefore

$$\int E_\nu d\sigma = (E + E_\gamma)\sigma + \int E' d\sigma(E, E_\gamma, P')d^3P'; \tag{6.17}$$

$d\sigma(E, E\ P')$ is simply the previous expression before the P' integration is performed, n_e is the Fermi electron density function, and n_γ is the Planck distribution.

The loss rate can be calculated in the four limits determined by relativity ($P \gg m$, $m \gg P$) and degeneracy ($E_F \gg kT$, $E_F \ll kT$). These rates are, in ergs/sec cc:

$-dE/dt$	conditions
$32\pi \times 10^6\, T_9^8$	non relativistic, non degenerate
$8\pi \times 10^{13}\, T_{10}^6 (\log_{10} T_{10} + 1.6)$	non relativistic, extreme degenerate
$20\pi \times 10^5 (1 + 5\, T_9^2)$	extreme relativistic, non degenerate
$(mc^2/E_F)^3\, T_9^7$	extreme relativistic, extreme degenerate

At a temperature of $T = 5 \times 10^8$ °K, during the period of carbon burning, the rate of energy loss is $10^{5.6}$ erg/g sec. The total available energy is 6 MeV/nucleon x 10^{23} nucleons/gm x 10^{-6} erg/MeV, about 6×10^{17} erg/gm. Therefore, the lifetime for ν decay is 10^{12} sec or 3×10^4 years. This is considerably shorter than the lifetime of the carbon burning stage computed without considering the neutrino flux.

If we look at massive stars in a cluster, their behavior on the H-R diagram looks like Figure 6.1.

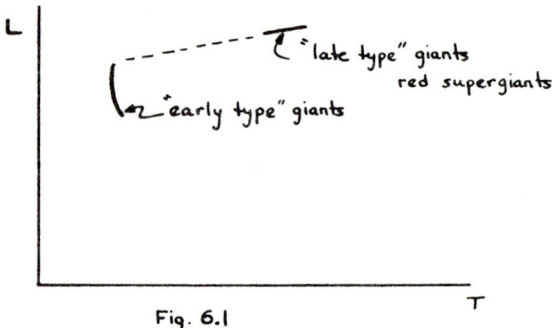

Fig. 6.1

This pattern is believed to reflect the evolution of these stars. They stay for considerable time in the region of "early type" giants, pass quickly through the dotted region and stay for a while in the region of red super giants. The number of stars in a given region should be roughly proportional to the length of time a star spends in that region.

According to the estimate of Hayashi and Cameron[*], neglecting neutrino emission, the star spends equal amounts of time as an early and a late giant, whereas with neutrino emission included the time spent as a late giant is fairly short. Consequently, the first theory predicts roughly equal numbers of stars in the two regions, while the latter requires fewer stars in the late region. At present the observational evidence cannot rule out either theory.

(d) The Pair Annihilation Process

The cross section can be increased by another power of $1/\alpha$, by considering a process with no photons. Such a process is the annihilation of an electron positron pair into neutrinos.

NEUTRINO ASTROPHYSICS

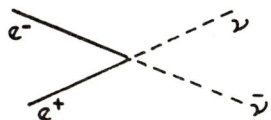

Again, the energy loss rate is given by

$$-dE/dt = \int \sigma v E_\nu \, N_{e^-} \, d^3 P_{e^-} \, N_{e^+} \, d^3 P_{e^+} \, . \qquad (6.18)$$

The pairs are going to be in thermodynamic equilibrium with the photons, for the time scale of pair production and pair annihilation to photons is much shorter than that of any neutrino process. The condition of thermodynamic equilibrium is that, if μ^-, μ^+, μ_γ are the chemical potentials for electrons, positrons and photons, respectively,

$$\mu^- + \mu^+ = \Sigma \mu_\gamma \, .$$

We write $\Sigma \mu_\gamma$ because the equilibrium reaction is $\gamma + \ldots + \gamma \rightleftharpoons e^+ + e^-$ with any number of photons participating.

The number of photons present at equilibrium is not constant, and is determined at a given temperature and pressure by

$$\partial G(N_\gamma)/\partial N_\gamma = 0 \, ; \qquad (6.19)$$

but

$$\partial G/\partial N_\gamma = \mu_\gamma \, ; \quad \therefore \mu_\gamma = 0 \, , \qquad (6.20)$$

so

$$\mu^- = -\mu^+ \equiv \mu \, . \qquad (6.21)$$

Then

$$N_{e^-} = \frac{2}{h^3} \left\{ \exp\left(\frac{E_- - \mu}{kT}\right) + 1 \right\}^{-1} ,$$

$$N_{e^+} = \frac{2}{h^3} \left\{ \exp\left(\frac{E_+ + \mu}{kT}\right) + 1 \right\}^{-1} . \qquad (6.22)$$

The probability of a reaction turns out to be, by calculating the Feynman diagram,

$$\frac{\sigma v}{c} = \frac{G^2 m^2}{2\pi} \left[\left(\frac{E_{e^-} + E_{e^+}}{m_e c^2}\right)^2 - 1 \right] \quad (6.23)$$

$$= 1.5 \times 10^{-45} \left[\left(\frac{E_{e^-} + E_{e^+}}{m_e c^2}\right)^2 - 1 \right].$$

If N_o is the number of electrons before the pairs are created, then by the principle of charge equality

$$n_{e^-} - n_{e^+} = N_o.$$

With many pairs, and ignoring for the moment N_o, we see that, first, the one in the denominator in Eq. (6.22) can be neglected, and second, since $n_{e^-} \approx n_{e^+}$, $\mu = 0$. Then,

$$n_{e^+} \approx n_{e^-} \sim \frac{2}{h^3} \int e^{-E/kT} d^3 P_e \quad (6.24)$$

$$= 8\pi \frac{m^3 c^3}{h^3} \int \exp\left\{-\sqrt{1+x^2}\; mc^2/kT\right\} x^2\, dx.$$

Since $mc/h \sim 1/\lambda_{Compton}$ this equation can be interpreted as saying that at $kT \sim mc^2$ one electron pair is produced in a volume $\sim$ the Compton wavelength cubed:

$$n_e \sim 10^{30} e^{-mc^2/kT} \quad (6.25)$$

The rate of energy production can now be estimated:

$$-dE/dt = \int E_\gamma \sigma v\, N_{e^+} N_{e^-}\, d^3 P_{e^+} d^3 P_{e^-} \quad (6.26)$$

$$\sim 10^{-5}\, \text{ergs}\; 10^{-45} \cdot 10^{10}\; 10^{30}\; 10^{30}\, e^{-2mc^2/kT};$$

at $T = 10^9$, $mc^2 \sim 6\, kT$; $e^{-2mc^2/kT} \sim e^{-12} \sim 10^{-5}$,

$$-dE/dt \sim 10^{15}\, \text{ergs/cm}^3\text{-sec}. \quad (6.27)$$

NEUTRINO ASTROPHYSICS

The energy density of photons at this temperature is

$$E_\gamma = aT^4 = 7.6 \times 10^{-15} (10^9)^4 = 10^{22} \text{ erg/cm}^3. \quad (6.28)$$

The lifetime for decay of photons to neutrinos is $\sim 10^7$ seconds.
Estimating the radius of the region which produces neutrinos to be 10^{10} cm, the neutrino luminosity is $10^{15} \times (10^{10})^3 \sim 10^{45}$ erg/sec. $L_\odot = 10^{33}$ erg/sec. Therefore $L_\nu = 10^{12} L_\odot$. It is interesting to note that if the sun were to undergo such a process, a man standing on earth would receive a radiation dose of 2 Roentgens from neutrinos alone!

We can find the temperature dependence of this reaction by noting that the important energy is $\sim kT$ and that for $T \gg T_o = mc^2/k$ the most important region in the integral

$$\left(\frac{mc}{h}\right)^3 \int_0^\infty \frac{x^2 dx}{e^{(\sqrt{1+x^2}\, T_o/T)} + 1}$$

is when $x^2 \gg 1$, i.e., when $\sqrt{1 + x^2} \approx x$.

Then we make the substitution $u = xT_o/T$, to obtain

$$n = 8\pi \left(\frac{mc}{h}\right)^3 \left(\frac{T}{T_o}\right)^3 \int_0^\infty \frac{u^2 \, du}{e^u + 1}. \quad (6.29)$$

The integral $\int_0^\infty \frac{u^2 \, du}{e^u + 1}$ is just a number, about 1.8.

Then, with $\sigma v = \sigma_o (4 E_T^2/mc^2 - 1)$, $E_T = E_+ + E_- \sim kT$.

We can write the energy loss as

$$-\frac{dE}{dt} = n_{e^-} n_{e^+} \langle \sigma v \rangle \langle E \rangle \sim T^3 T^3 T^2 T = T^9. \quad (6.30)$$

In fact,

$$-\frac{dE}{dt} \approx AT^9 \approx 4 \times 10^{15} T_9^9 \text{ ergs/cc-sec}, \quad (6.31)$$

valid to 10% at $T = 5 \times 10^9$.

(e) Plasma Process

A free photon cannot decay into neutrinos because the decay of a zero mass particle into two particles is forbidden by energy-momentum conservation. In an electron gas photons may appear to have a rest mass. The relation between frequency ω and wave vector k is, for $\omega > \omega_o$ (ω_o the plasma frequency)

$$\hbar^2\omega^2 = \hbar^2\omega_o^2 + k^2c^2 . \qquad (6.32)$$

In this case a photon may decay into **a neutrino**-antineutrino pair. To evaluate the energy generation, one must quantize such a plasma wave and evaluate ω_o for a relativistic degenerate electron gas.

The plasma frequency ω_o is related to the dielectric constant approximately by

$$\epsilon \approx 1 - \omega^2/\omega_o^2 ; \qquad (6.33)$$

ϵ, and consequently ω_o, can be calculated from dispersion relations. The real part of ϵ is related to the imaginary or absorptive part by the familiar relation:

$$R_e [\epsilon(\omega) - 1] = \frac{2}{\pi} \int \frac{\omega' \mathcal{Im}\, \epsilon(\omega')}{\omega'^2 - \omega^2 - i\delta} d\omega', \quad \delta > 0 . \qquad (6.34)$$

The absorptive part is evaluated by explicit calculation of the processes in which a photon can be absorbed. The result is

$$\omega_o^2 = \frac{4}{3\hbar^3} e^3 P_F^3 \frac{1}{\pi E_F} ; \qquad (6.35)$$

P_F and E_F are the Fermi energy and momentum.

The quantization of the field turns out to be similar to that of a free photon, but with different normalization.

Without discussing details, I will give the results. The reaction proceeds through the diagram:

$$-\frac{dE}{dt} = 1.23 \times 10^{22} \left(\frac{\hbar\omega_o}{mc^2}\right)^9 \sum_{n=1}^{\infty} \frac{K_2(n\hbar\omega_o/kT)}{n\hbar\omega_o/kT} \frac{\text{ergs}}{\text{cc-sec}} ;$$

$$-\frac{dE}{dt} = 2.96 \times 10^{22} \left(\frac{\hbar\omega_o}{mc^2}\right)^6 \left(\frac{mc^2}{kT}\right)^{-3} \frac{\text{ergs}}{\text{cc-sec}}, \quad \hbar\omega_o \ll kT; \quad (6.36)$$

$$-\frac{dE}{dt} = 1.54 \times 10^{22} \left(\frac{\hbar\omega_o}{mc^2}\right)^{7.5} \left(\frac{mc^2}{kT}\right)^{-1.5} e^{-\hbar\omega_o/kT}, \quad (6.37)$$
$$\hbar\omega_o \gg kT.$$

It can be seen that this process has a sharp maximum at $\omega_o \sim kT/\hbar$. It is very effective in degenerate matter, say, at $\rho \sim 10^6$ g/cc, $T \sim 10^8$ °K, values appropriate to the core of a helium-burning red giant, when other processes are unimportant.

(f) Summary

The pair annihilation process is most effective at temperatures above 10^9 °K. The plasma process is most effective at high density at $T < 10^9$ °K. In the intermediate region of the ρ-T plane (see Fig. 6.2), the photo-neutrino process dominates for all temperatures above $10^{7.5}$ °K.

Thus the pair annihilation process is the one most likely responsible for the neutrino process of stellar collapse, and the plasma process is most likely responsible for the creation of white dwarfs.*

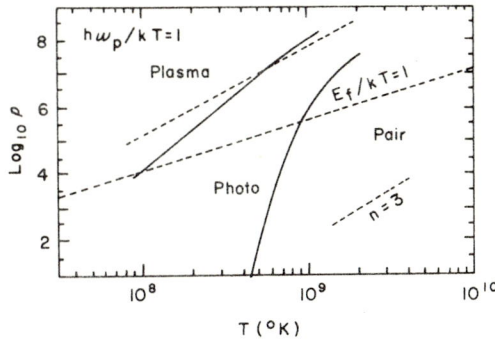

Fig. 6.2. Dominant mechanisms of energy loss for various densities and temperatures.

---

* The first results of numerical integration show that the structure of stars when neutrino processes are included may be substantially different from the models with the neutrino processes neglected. Consequently all the discussion in this chapter should be considered tentative, pending more detailed numerical results.

Energy loss rates are summarized in Figs. 6.3-6.7.

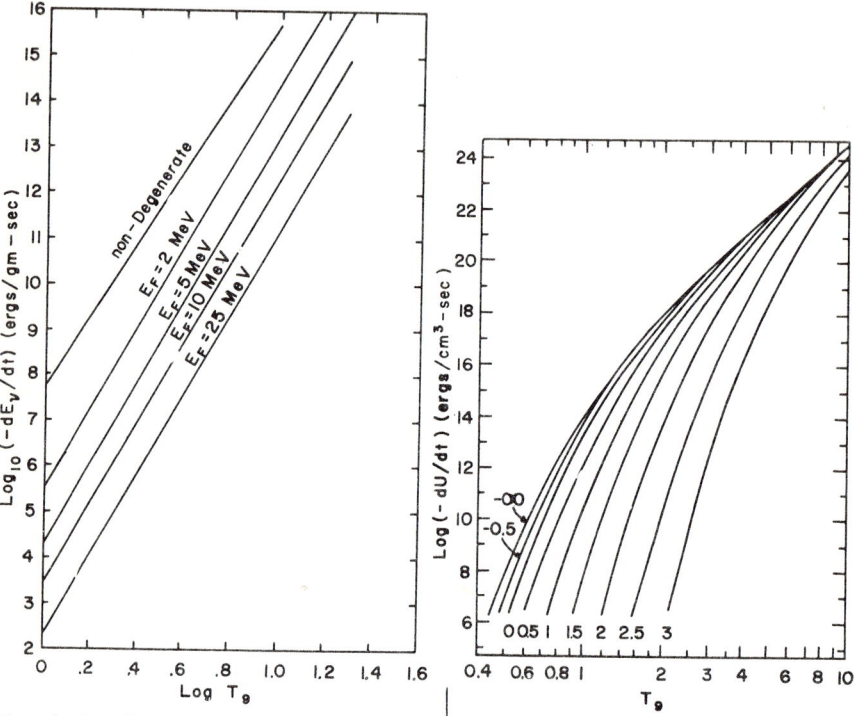

Fig. 6.3. Energy loss rates for the photo-neutrino process (Chiu and Stabler). The energy loss rate per unit mass is plotted against the temperature for several values of the Fermi energy. The relation of density ρ to Fermi energy is roughly $\rho = 1.6 \times 10^6 (E_F/\text{MeV})^3$, when $E_F \gg 1/2$ MeV.

Fig. 6.4. Energy loss rates for the annihilation process are plotted as a function of temperature for several values of density. Numbers on the curves refer to $\log_{10}(\rho/\rho_0)$, where $\rho_0 = 6 \times 10^6$ g/cm^3.

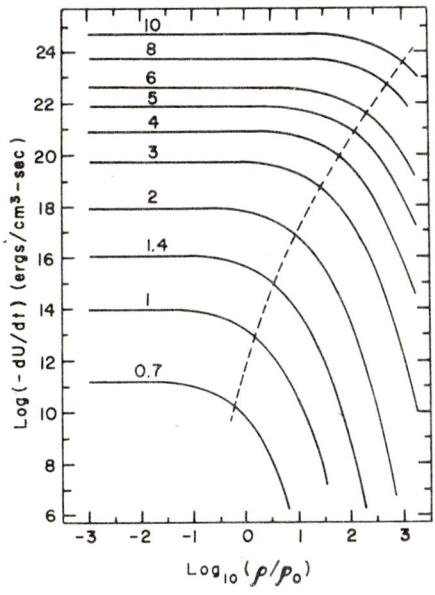

Fig. 6.5. Energy loss rates for the annihilation process are plotted as a function of $\log_{10}(\rho/\rho_0)$, where $\rho_0 = 6 \times 10^6$ g/cm^3. Numbers on the curve refer to temperature in units of 10^9 °K.

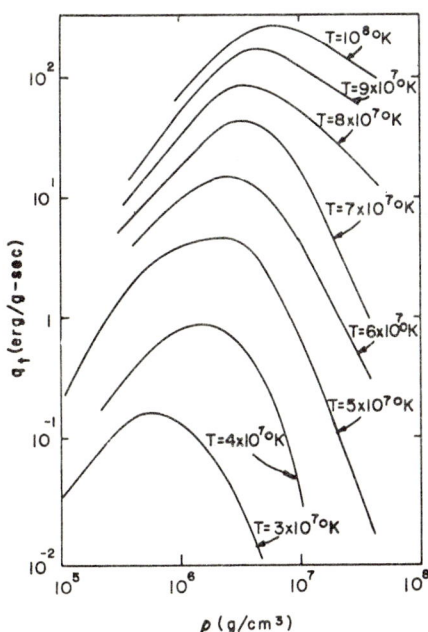

Fig. 6.6. Energy loss rates for the plasma process are plotted as a function of density for small values of temperature. (Adams, Ruderman, and Woo).

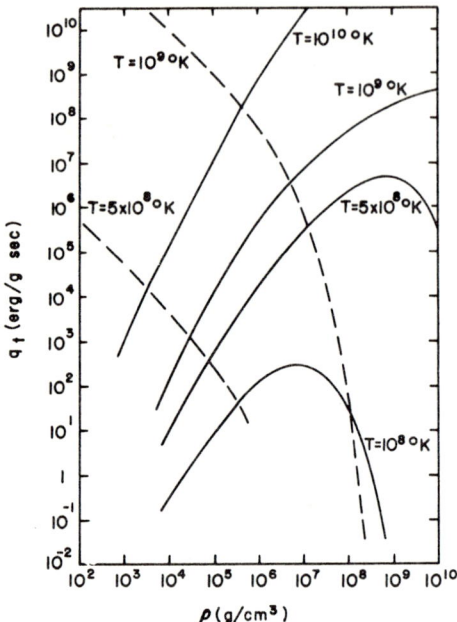

Fig. 6.7. Energy loss rates for the plasma process for higher values of temperature (Adams, Ruderman and Woo).

6.2 Stellar Collapse

(a) Neutrino Process

We have discussed the neutrino processes responsible for rapid energy loss from the star. This energy loss can lead to collapse of the star. There are actually two processes that can cause stellar collapse. First, we consider the neutrino process.

Recall the equilibrium equations for a star in radiative equilibrium:

$$\frac{dP}{dr} = -\rho \frac{GM}{r^2} , \qquad (6.38)$$

NEUTRINO ASTROPHYSICS

$$\frac{dm}{dr} = 4\pi r^2 \rho , \tag{6.39}$$

$$\frac{dL}{dr} = 4\pi r^2 \rho \epsilon , \tag{6.40}$$

$$T^3 \frac{dT}{dr} = \frac{1}{\text{const}\,\rho} \frac{L}{4\pi r^2} \tag{6.41}$$

Let us see how these equations are changed.
First, if the star is collapsing, then matter is accelerating inward. The gravitational force is not balanced by the pressure force, and the first equation is modified to

$$\frac{dP}{dr} = -\rho \frac{GM}{r^2} - \rho \ddot{r} . \tag{6.42}$$

We could consider a point in space, in which case

$$\ddot{r} = \frac{\partial \vec{v}}{\partial t} + \vec{v} \cdot \vec{\nabla} \vec{v} , \tag{6.43}$$

but it is simpler to use m as the independent variable since m is independent of time. That is, we follow a particular shell as it falls in:

$$\frac{dP}{dr} = \frac{dP}{dm}\frac{dm}{dr} = 4\pi r^2 \rho \frac{dP}{dm} . \tag{6.44}$$

The equations are

$$\frac{dP}{dm} = -\frac{Gm}{4\pi r^4} - \frac{\ddot{r}}{4\pi r^2} , \tag{6.45}$$

$$\frac{dr}{dm} = \frac{1}{4\pi r^2 \rho} , \tag{6.46}$$

$$\frac{dL}{dm} = \epsilon . \tag{6.47}$$

The equation of radiative equilibrium is now no longer coupled to the other equations, since $L \approx L_\nu \gg L_\gamma$.

The last equation should be replaced by

$$-T \frac{dS}{dt} = \epsilon_\nu(\rho, T) , \qquad (6.48)$$

where S is the entropy per unit mass. For then with the thermodynamic equations

$$S = S(\rho, T) , \\ P = P(\rho, T) , \qquad (6.49)$$

and the continuity equation

$$d\rho/dt + \bar{v} \cdot \bar{\nabla}\rho = 0 , \qquad (6.50)$$

the set of equations is complete and soluble for the four unknowns (ρ, T, r, v) in terms of the independent variables m and t.

The total energy W, because of the Virial Theorem, is within a constant factor of the order of unity

$$W \approx - MR_g T . \qquad (6.51)$$

Then,

$$\frac{dW}{dt} = + MR_g \frac{dT}{dt} , \qquad (6.52)$$

and by the results of neutrino processes,

$$\frac{dW}{dt} = M \frac{AT^9}{\rho} . \qquad (6.53)$$

Because $MR \approx \frac{2GM^2}{R_g}$

and $\rho \approx M / \frac{4}{3}\pi R^3$,

we conclude first,

$$\rho = [\frac{R_g^3}{32G^3 M^2}] T^3 \equiv aT^3 ; \qquad (6.54)$$

also

NEUTRINO ASTROPHYSICS

$$\frac{dR}{dt} = -\frac{2GM}{R_g}\frac{1}{T^2}\frac{dT}{dt} = -\frac{2GM}{R_g}\frac{AT^4}{a} \qquad (6.55)$$

$$= -\frac{C}{R^4}$$

and

$$\frac{d^2R}{dt^2} = -\frac{4C^2}{R^9} \qquad (6.56)$$

Since both dR/dt and d^2R/dt^2 are negative, the star must be collapsing inward with increasing speed.

The acceleration can never be greater than gravitational acceleration; therefore

$$\left|\frac{d^2R}{dt^2}\right| = \frac{4C^2}{R^9} < \frac{MG}{R^2} \qquad (6.57)$$

This gives a limiting radius R_ℓ below which hydrostatic equilibrium is impossible:

$$R_\ell = \left(\frac{4C^2}{MG}\right)^{1/7}. \qquad (6.58)$$

For $M = 20\ M_\odot$

$R_\ell \approx 10^9$

$C = 1.08 \times 10^{31}$

$T \approx 4 \times 10^9\ °K$.

Now for T less than this, $kT \ll E_F$, so this temperature is not much above what would effectively be cold matter. The mass limit for cold matter is 1.44 $M_\odot$ (Chandrasekhar limit) for a white dwarf, or as we discuss below, only 0.8 $M_\odot$ for a neutron star. So there is a problem as to what to do with so much matter.

(b) Fe-He Phase Transition

There is a second possible cause of stellar collapse. The criterion for stable equilibrium we derived in Chapter 2 was

$$\int (\Gamma - \tfrac{4}{3}) dV > 0 . \qquad (6.59)$$

As nucleosynthesis proceeds, the proportion of Fe in the star builds up. But at about $8 \times 10^9\ °K$ statistical equilibrium favors He over Fe. The conversion of Fe to He has a coefficient $\Gamma < 4/3$ and can cause instability in the equilibrium, leading to collapse.

We previously derived the equation

$$\Gamma = -\frac{v}{P} \frac{(P_x \epsilon_y - \epsilon_x P_y) + P(P_x v_y - P_y v_x)}{v_x \epsilon_y - v_y \epsilon_x} . \qquad (6.60)$$

For $x = P$, $y = v$,

$$\Gamma = \frac{Pv + v \frac{\partial \epsilon}{\partial v}}{P \frac{\partial \epsilon}{\partial P}} . \qquad (6.61)$$

This can be put in the form

$$\Gamma = 1 + \frac{\Delta(Pv)}{\Delta(E)} . \qquad (6.62)$$

For example, for a perfect gas $\epsilon(T) = 3/2\ R_g T$,

$$Pv(T) = R_g T + a T^4 .$$

For radiation, $Pv = 1/3 \epsilon$, so

$$\Gamma = 1 + 1/3 = 4/3 . \qquad (6.63)$$

For the iron-helium transition the motion is relativistic so $Pv = 1/3\ E_{kin}$, and $\epsilon = E_{kin} + E_{transition}$;

$$E_{kinetic} = 1\ MeV,$$

$$E_{trans} = 2.2\ MeV,$$

$$\Gamma = 1 + \frac{1/3}{1 + 2.2} = 1.11 < 4/3 . \qquad (6.64)$$

It is not known which of these two processes is working in stellar collapse.

NEUTRINO ASTROPHYSICS

The core of the star contains no fuel and consequently no nuclear energy is produced in the core during collapse. Fuel remains in the envelope, however, with the energy generation rate depending on temperature like $\epsilon = \epsilon_o T^{26}$. Consequently the envelope is explosive. Carbon fuel is used up in about 10 seconds. A high density of fast neutrons and protons build up many heavy elements during the supernova explosion. As a result of the imploding core and exploding envelope, there is a discontinuity, which engenders a shock wave. At the surface the shock wave propagates nearly with the velocity c, and kicks particles out to space with energy up to a BeV. This may start part of the cosmic ray spectrum on their way, although by the time cosmic rays reach earth much of the spectrum has significantly higher energy.

Several objects in the sky are believed to be remnants of supernovae. Occasionally extragalactic supernovae may be observed. For a time it was believed that the light curve of supernovae followed a law

$$L = L_o e^{-t/55 \text{ days}} . \qquad (6.65)$$

The observation of Cf^{254}, which has a 55-day decay law, in the H-bomb explosion of 1952, prompted the hypothesis that the light was due to Cf^{254}. But it would take a lot of Cf^{254} to produce so much light. It has since been realized that the light curve is a superposition of many curves, in particular a 77-day curve and a 45-day curve.

7. INTRODUCTION TO GENERAL RELATIVITY

We are going to relate cosmology to various things. Primarily we want to find out what limits are put on certain ideas in physics by other, basic, principles.

For example, to what extent can we talk about the gravitational energy of a proton?

The uncertainty principle says

$$\Delta E \Delta t \gtrsim \hbar .$$

The life of the universe is $\Delta t \sim 10^{18}$ sec. Therefore ΔE, the precision with which we can specify the mass of the proton throughout the life of the universe is

$$\Delta E \sim 10^{-27}\, 10^{-18} \sim 10^{-45} \text{ ergs} .$$

The self-gravitational energy of the proton is

$$\frac{Gm^2}{r} \doteq \frac{10^{-7}(10^{-24})^2}{10^{-13}} = 10^{-42} \text{ ergs},$$

not much larger than the limit of precision set by the uncertainty principle. The notion of the self-gravitational energy of a single proton may well have no meaning at all.

As another example, Wigner pointed out that a microscopic time measurement must nevertheless require a macroscopic clock to measure it. For example, suppose we wish to measure time to within 10^{-8} sec in a running time of one day, 10^5 seconds. This is an accuracy of $1:10^{-13}$. To measure this we need a clock with 10^{13} distinct states at least. For, suppose that time were measured by a ball jumping between two pots once for each unit of time. If more than two units of time elapse, we need another set of pots to measure how many times the ball has gone back and forth. Wigner calculates that the clock above would require at least a gram of matter.

We also point out that in a macroscopic time, light travels a macroscopic distance, say 3×10^{-3} cm in 10^{-13} sec.

In this last example, the requirements of quantum mechanics, together with the requirement that we specify explicitly how we are going to measure time, puts limits on our ability to do it.

Now we summarize the principles of general relativity to see what limitations they put on our understanding of physics. The first two are general, and are closely related:
1. All statements in physics must have an operational interpretation.
2. The basis of physical theories is experimental.

The next three are specifically related to relativity, and the experimental verification of relativity constitutes the grounds on which they may be legitimately postulated:
3. A four-dimensional geometry is necessary.
4. The principle of equivalence.
5. The principle of covariance.

The need for a four-dimensional geometry is related to the lack of simultaneity of two events seen by one observer if another observer moving past him does see the events as simultaneous. This can be illustrated by means of a diagram, Fig. 7.1.

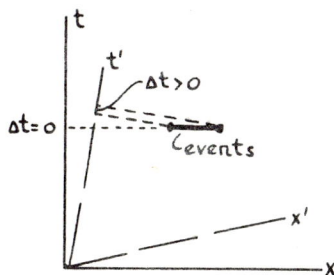

Fig. 7.1. The Relativity of Simultaneity. The events simultaneous in (x,t), $\Delta t = 0$, are not simultaneous in the Lorentz rotated coordinate system (x',t').

Prior to the question of whether there is four-dimensional geometry is the question of whether geometry is an appropriate description at all. This will be discussed below.

The principle of equivalence was stated by Einstein as "The laws of physics must be of such a nature that they apply to systems of reference in any kind of motion." This tacitly assumes that one knows how to transform any laws of physics that have been or will be discovered.

Dicke has stated this principle in two forms which have direct operational meaning. The weak principle states the local equivalence of gravitational and inertial effects:

> All material particles in motion (that is, test particles whose mass is negligible compared to the source of the gravitational field) follow the same trajectories in space-time, provided the initial conditions (position and velocity) are the same.

The strong principle asserts that physics laws are homogeneous in space and time. This means that in a freely falling laboratory in any space-time locality, one will get the same experimental result as in any other providing that the change in gravitation over the space-time distance in which the experiment is performed is sufficiently small to satisfy your criterion of accuracy.

Note that even a satellite is not necessarily a laboratory which adequately satisfies the criterion of the strong principle. As Fig. 7.2 illustrates, because the gravitational potential depends on distance, the force on an object at one end of the satellite is greater than on the other. If an experiments lasts sufficiently long, this can become important.

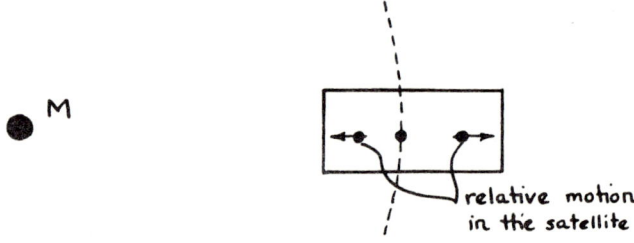

Fig. 7.2

The relative strengths of the fundamental interactions are:

strong 1

electromagnetic 10^{-2}

weak 10^{-12}

gravitational 10^{-40} proton

10^{-29} gram of matter

10^{-6} sun

10^{-8} Jupiter

NEUTRINO ASTROPHYSICS

The principle of equivalence asserts the constancy of these ratios locally (weak) or globally (strong) for contributions to inertial as compared to gravitational mass.

The Eötvös experiment is designed to check the principle of equivalence. Essentially the experiment involves a swinging mass. The ratio of gravitational to centrifugal force is compared for different materials. Eötvös, in 1920, found the ratio constant within one part in 10^9. Dicke, in 1961, repeating the experiment, found constancy to within 5 parts in 10^{12}.

The Eötvös experiment excludes the possibility that anti-particles anti-gravitate. A comparison of the virtual positron density in heavy metals as compared to that in light materials indicates that an effect of one part in 10^8 would be expected, which is not seen.

Einstein's statement of the principle of covariance is:
"The general laws of nature are to be expressed by equations which hold good for all systems of coordinates; that is, are co-variant with respect to any substitutions whatever."

An empty interpretation of this statement is that the physics is independent of the curved lines you draw in space. The content of the statement is that physics is not restricted to flat geometries. In a more general interpretation the geometry of space-time is determined from the physics, and not given.

In any system, we can transform the metric tensor g_{ij} to the flat space form at a point

$$g_{ij}(x_o) = \begin{pmatrix} 1 & & & \\ & 1 & & \\ & & 1 & \\ & & & -1 \end{pmatrix} \equiv \eta_{ij} .$$

Except for the case of flat space, this is only true at one point. Therefore we may write

$$g_{ij}(x) = \eta_{ij} + h_{ij}(x), \qquad h_{ij}(x_o) = 0 ,$$

$$h_{ij}(x) \neq 0 \quad x \neq x_o .$$

The relevance of writing physical laws in a way which is independent of the metric depends on whether g_{ij} differs appreciably from η_{ij} over the region of interest. Changes in g_{ij} are governed by the gravitational field equation. For atomic physics Δx, the region of interest, is $\sim 10^{-8}$ cm. For nuclear physics, $\Delta x \sim 10^{-13}$ cm. Not even in neutron stars does g_{ij} change appreciably over such small distances. Consequently co-variance is probably irrelevent in atomic and elementary particle physics.

Does general relativity satisfy the principle of co-variance, or is there a preferred co-ordinate system? The field equation is

$$R_{\mu\nu} - \frac{1}{2} g_{\mu\nu} R = T_{\mu\nu}.$$

The law of conservation of energy requires $T_{\mu\nu;\nu} = 0$. $A;\nu$ indicates the co-variant derivative of A in the ν direction.

There are ten quantities $g_{\mu\nu}$ and ten equations. In order that $g_{\mu\nu}$ be undetermined so that it admits four co-ordinate transformations, only six of the ten equations for $g_{\mu\nu}$ can be independent and non-trivial. This is guaranteed because the combination $R_{\mu\nu} - 1/2\, g_{\mu\nu}R$ satisfies the equation

$$(R_{\mu\nu} - \frac{1}{2} g_{\mu\nu} R)_{;\nu} \equiv 0 \quad (= T_{\mu\nu,\nu})$$

trivially, that is, identically. The constant 1/2 was chosen for this reason. Four of the ten equations contain no information about $g_{\mu\nu}$.

The field equations are constructed so that the physical information carried by $g_{\mu\nu}$ is independent of the co-ordinates. Yet, to solve any particular problem a coordinate system must be chosen and the $g_{\mu\nu}$ specified explicitly. There will be four non-covariant equations. These are called co-ordinate conditions.

Is any set of co-ordinate conditions preferred? Fock has argued that the "harmonic coordinates" defined by

$$(\sqrt{-\det g'} \; g^{\mu\nu})_{,\nu} = 0$$

together with boundary conditions at infinity, is a preferred system, because it is uniquely determined up to a Lorentz transformation. Others hold the choice of co-ordinates to be completely arbitrary. I believe that at this point the question of preferred co-ordinates is one of taste, not of physics anymore.

Finally, let us consider the experimental evidence for three ideas which are often taken for granted:
1. Isotropy of mass;
2. Constancy of the electric charge independent of motion;
3. Justification of using geometry to describe space and time.

Mach's principle asserts that the inertial properties of matter are determined by the properties of matter in the universe. There are two local sources of anisotropy of matter -- the sun and the galaxy. Can they cause an observable anisotropy of mass? Such anisotropy would show up as a frequency shift in dipole transitions when the apparatus is rotated relative to the galaxy and sun. No effects were

NEUTRINO ASTROPHYSICS 73

observed within an accuracy of one part in 10^{10}. This is approximately the accuracy that would be needed to detect the effects expected assuming a reasonable mechanism for the anisotropy to occur. Using the Mössbauer effect and nuclear magnetic resonance of Li^6, the degree of isotropy has been established to one part in 10^{22}.

Dicke asserted that Mach's principle requires that anisotropies be unobservable because all instruments have the same anisotropy.

An experiment was devised to measure the difference in charge between the proton and electron. As a by-product the velocity dependence of the electric charge is tested.

A beam of atoms is sent between two charged plates and detected 10 meters away. If there is a net charge on the atom, it will be deflected. Because the electron is in motion in the atom, if there is no deflection, we may conclude the electron and proton charges are equally independent of velocity. The result of this experiment is that

$$\frac{e_p - e_e}{e_p} < 10^{-22}$$

and if $e_e = e_o + A \frac{v^2}{c^2}$, then $A < 10^{-18}$.

Suppose we moved a measuring rod from some point A to some other point and brought it back. In order that its length L at A be an invariant, we require that it have the same length after it returns as it had previously. Its length depends on the value of the Bohr radius.

$$r_a = \hbar^2 / m_e e^2.$$

Assuming the strong principle of equivalence, $\hbar$ and e are constants of nature. Therefore, any inconsistency in the unit of length would show up as a difference in mass of electrons with different histories of travel in the universe.

The Pauli principle declares that identical fermions cannot occupy the same quantum state. If two electrons had different mass, the exclusion principle would not apply.

In any atom, any electrons with different mass than those in the ground state would collapse to the ground state. The existence of atoms is evidence that the mass of electrons is constant within the accuracy that the uncertainty principle allows. The uncertainty ΔE is related to the time that the inner shell electrons have undergone no transitions. Let $t = 10^6$ years (a time short compared to the age of the universe). Then,

$$\Delta E = \hbar / t$$

$$\Delta E = 10^{-40} \text{ ergs}$$

and $$\Delta E / mc^2 = 10^{-34}.$$

This is a crude measure of the extent to which we can say the use of geometry is experimentally justified.

The limits of accuracy to which these three ideas have been experimentally tested gives the limits to which all physical theories must incorporate them. Moreover, since experimental verification is the foundation of all physical theory, the validity of no idea is established until there is experimental support for it.

8. NEUTRON STARS

For the discussion of neutron stars, some knowledge of general relativity is necessary. We will not discuss this here, because of the wealth of information in available textbooks. See the Bibliography.

Einstein's field equation is

$$R_{\mu\nu} - \frac{1}{2} g_{\mu\nu} R = \frac{8\pi G}{c^4} T_{\mu\nu} ; \qquad (8.1)$$

$$R_{\mu\nu} = B^i_{\mu\nu i} ,$$

where $B^{\alpha}_{\rho\gamma\delta}$ is the "key tensor,"

$$B^i_{\kappa\lambda\mu} = \Gamma^i_{\kappa\lambda,\mu} + \Gamma^i_{\kappa\mu,\lambda} + \Gamma^i_{\alpha\lambda}\Gamma^{\alpha}_{\kappa\mu} - \Gamma^i_{\alpha\mu}\Gamma^{\alpha}_{\kappa\lambda} . \qquad (8.2)$$

$\Gamma^{\sigma}_{\mu\nu}$ is in our space equivalently the symmetric affine connection or the Christoffel symbol,

$$\Gamma^{\sigma}_{\mu\nu} = \frac{1}{2} g^{\sigma\lambda} \left(\frac{\partial g_{\mu\nu}}{\partial x^{\nu}} + \frac{\partial g_{\nu\lambda}}{\partial x^{\mu}} - \frac{\partial g_{\mu\nu}}{\partial x^{\lambda}} \right) ; \qquad (8.3)$$

$g_{\mu\nu}$ is the metric tensor,

$$ds^2 = g_{\mu\nu} dx^{\mu} dx^{\nu} \qquad (8.4)$$

and $\quad g^{\mu\nu} = (g_{\mu\nu})^{-1} , \quad g^{\mu\alpha} g_{\alpha\nu} = \delta^{\mu}_{\nu} = \begin{cases} 1 & \mu = \nu \\ 0 & \mu \neq \nu \end{cases} .$

There is no general solution of these equations for arbitrary energy tensor $T_{\mu\nu}$.

8.1 Schwarzschild Exterior Solution

Schwarzschild found a particular solution for the case which corresponds to the Newtonian equation

$$\nabla^2 \phi = -4\pi \rho(r).$$

The assumptions made are:
1. Spherical symmetry, and no time dependence. That is, the metric is a function only of r.
2. No rotation.
3. The solution reduces to Newtonian form at large distances.

The "three famous tests" of general relativity (the advance of the perihelion of mercury, the bending of starlight, and the gravitational red shift) all were based on the Schwarzschild solution.

First, let us assume the metric defined by

$$ds^2 = -e^{\lambda'} dr^2 - e^{\mu'}(r^2 d\theta^2 + r^2 \sin^2\theta \, d\phi^2) + e^{\nu'} dt^2 + 2a' dr\, dt, \quad (8.5)$$

$\lambda' = \lambda'(r)$, $\mu' = \mu'(r)$, $\nu' = \nu'(r)$, $a' = a'(r)$. (The primes do not indicate time independent differentiation.)

This is the most general, spherically symmetric, metric.

We now make a co-ordinate transformation $R = \left[e^{1/2 \mu'} \right] r$.

Then,

$$ds^2 = e^{\lambda'} f^2(\mu') dR^2 - R^2(d\theta^2 + \sin^2\theta \, d\phi^2) + e^{\nu'} dt^2 + 2a'' dR\, dt, \quad (8.6)$$

where $f^2(\mu') = \left\{ e^{-\frac{1}{2}\mu'(r)} - \frac{1}{2} r \left[\frac{\partial}{\partial r}\mu'(r)\right] e^{-\frac{1}{2}\mu'(r)} \right\}^2$

and $\quad a'' = a' \left\{ f(\mu') \right\}$.

Now we write

$$e^{\nu'} dt^2 + 2a'' dt\, dr = (e^{\frac{1}{2}\nu'} dt + a'' dR)^2 - (a'')^2 dr^2. \quad (8.7)$$

Because it is composed of two terms, we can always find an integrating factor η so that

$$e^{1/2 \nu'} dt + a'' dR = \frac{1}{\eta} d\tau. \quad (8.8)$$

Now we have

NEUTRINO ASTROPHYSICS

$$ds^2 = -(e^{\lambda'} f(\mu') + a''^2) dR^2 - R^2(d\theta^2 + \sin^2\theta \, d\phi^2) + \frac{1}{\eta^2} d\tau^2. \quad (8.9)$$

For convenience, we make the substitutions

$$e^{\lambda'(r)} f(\mu'(r)) + a''^2(r) \rightarrow e^{\lambda(R)},$$

$$\frac{1}{\eta^2} \rightarrow e^{\nu(R)},$$

$$\tau \rightarrow t,$$

$$R \rightarrow r.$$

Then

$$ds^2 = -e^\lambda dr^2 - r^2 d\theta^2 - r^2 \sin^2\theta \, d\phi^2 + e^\nu dt^2 \quad (8.10a)$$

or

$$g_{\mu\nu} = \begin{pmatrix} -e^\lambda & & & \\ & -r^2 & & \\ & & -r^2 \sin^2\theta & \\ & & & e^\nu \end{pmatrix} \quad (8.10b)$$

Since the metric tensor is diagonal,

$$g^{\mu\nu} = -1/g_{\mu\nu}.$$

To illustrate the calculation of the Γ's

$$\Gamma^1_{11} = \frac{1}{2} g^{1\lambda} \left(\frac{\partial g_{1\lambda}}{\partial x^1} + \frac{\partial g_{1\lambda}}{\partial x^1} - \frac{\partial g_{11}}{\partial x^\lambda} \right). \quad (8.11)$$

Since $g_{\mu\nu}$ is diagonal, $g_{1\lambda} = 0$ unless $\lambda = 1$. So

$$\Gamma^1_{11} = \frac{1}{2} g^{11} \left[\frac{\partial g_{11}}{\partial x_1} + \left(\frac{\partial g_{11}}{\partial x_1} - \frac{\partial g_{11}}{\partial x_1} \right) \right] \quad (8.12)$$

$$= \frac{1}{2} e^{-\lambda(r)} \left(\frac{d\lambda(r)}{dr} e^{\lambda(r)} \right) = \frac{1}{2} \frac{d}{dr} (\lambda(r)).$$

We list below all the non-zero Γ. *

---

*From this point on primes indicate differentiation with respect to r.

$$\Gamma_{11}^1 = 1/2\,\lambda' \qquad \Gamma_{12}^2 = 1/r$$

$$\Gamma_{22}^1 = -re^{-\lambda} \qquad \Gamma_{13}^3 = 1/r$$

$$\Gamma_{33}^1 = -re^\lambda \sin^2\theta \qquad \Gamma_{14}^4 = 1/2\,\nu' \qquad (8.13)$$

$$\Gamma_{44}^1 = 1/2\,\nu' e^{\nu-\lambda} \qquad \Gamma_{33}^2 = -\sin\theta\cos\theta$$

$$\Gamma_{22}^3 = \cot\theta$$

and
$$\Gamma_{\lambda\kappa}^\mu = \Gamma_{\kappa\lambda}^\mu \,.$$

A long, but straightforward, calculation gives

i) $\;-8\pi T_{11} = e^{-\lambda}(\nu'/r + 1/r^2) - 1/r^2 \;;$

ii) $\;-8\pi T_{22} = -8\pi T_{33} = e^{-\lambda}(\nu''/2 - \lambda'\nu'/4 + \nu'^2/4 + (\nu'-\lambda')/2r);\quad (8.14)$

iii) $\;-8\pi T_{44} = e^{-\lambda}(-\lambda'/r + 1/r^2) - 1/r^2 \,.$

The units are given by $c = G = 1$.
  Outside all matter $T_{\mu\nu} = 0$. Then, equating Eq. (8.14)(i) to (8.14)(III),

$$\lambda' = -\nu' , \qquad (8.15)$$

and Eq. (8.14)(ii) becomes

$$\nu'' + \nu'^2 + 2\nu'/r = 0 \,. \qquad (8.16)$$

NEUTRINO ASTROPHYSICS

This equation can be integrated; or we can integrate Eq. (8.14)(iii) to give

$$e^\nu = a + b/r, \quad e^{-\lambda} = A + B/r. \tag{8.17}$$

To determine the four constants, first we require that the metric go to the flat space form:

$$ds^2 \xrightarrow[r \to \infty]{} -(dr^2 + r^2 d\theta^2 + r^2 \sin^2\theta d\phi^2) + dt^2.$$

This determines

$$A = a = 1.$$

An analysis of the Newtonian case shows that in the weak field limit $g_{44} = 1 - 2m/r$.

This, and $-\nu' = \lambda'$, so that $e^\nu = e^{\text{const}} e^\lambda$, with $a = A = 1$, so that const $= 0$, determines the metric tensor finally as

$$g_{\mu\nu} = \begin{pmatrix} -\dfrac{1}{1-\dfrac{2m}{r}} & & & \\ & -r^2 & & \\ & & -r^2 \sin^2\theta & \\ & & & 1-\dfrac{2m}{r} \end{pmatrix} \tag{8.18a}$$

or

$$ds^2 = -\dfrac{dr^2}{1-\dfrac{2m}{r}} - r^2 d\theta^2 - r^2 \sin^2\theta d\phi^2 + (1 - 2m/r)dt^2. \tag{8.18b}$$

The metric ds^2 seems to have a singularity at $r = 2m$. Does this singularity have physical significance?

Restoring G and c, we see that at this radius

$$1 = \dfrac{2m}{r} = \dfrac{2GM}{rmc^2} \quad (m \text{ is the mass of the test particle})$$

$$= \dfrac{2GM}{r}/Mc^2. \tag{8.19}$$

This means that at this radius, the "Schwarzschild radius," the gravitational self-energy of the body is the same as it's rest energy. Or, according to the first equation, since $2GMm/r$ is the energy needed for escape, a body of mass m would have to convert all its self energy to propel itself ballistically away.

The singularity in the metric can be transformed away by making a co-ordinate transformation. For example:

$$r = (1 + \frac{m}{2\bar{r}})^2 \bar{r} . \qquad (8.20)$$

The new metric is

$$ds^2 = -(1 + \frac{m}{2\bar{r}})^4 (d\bar{r}^2 + \bar{r}^2 d\theta^2 + \bar{r}^2 \sin^2\theta d\phi^2) + (\frac{1 - m/2\bar{r}}{1 + m/2\bar{r}}) dt^2 . \qquad (8.21)$$

This is called the isotropic co-ordinate system. It is the co-ordinate system on a falling body. In isotropic co-ordinates there is no singularity in the metric.

Let us look more closely at the Schwarzschild radius $r_s = 2GM/c^2$.

For a proton:

$$r_s = \frac{2 \times 6 \times 10^{-8} \times 1.7 \times 10^{-24}}{(3 \times 10^{10})^2} = 10^{-52} \text{ cm} . \qquad (8.22)$$

For the sun:

$$r_s = 1.3 \times 10^{-28} \times 2 \times 10^{33} = 2.6 \text{ km.} \qquad (8.23)$$

The density of the sun, if it were all collapsed inside the Schwarzschild radius, would be

$$\frac{2 \times 10^{33}}{\frac{4}{3} \times 27 \times 10^{15}} \sim 10^{16} \text{ g/cc}$$

This gives the order of magnitude of density for which relativity becomes very important inside a star.

The galaxy has a mass of about 10^{10} suns; $M_g = 10^{43}$ gm. The universe has about 10^{10} galaxies. This gives $M_u = 10^{53}$ gms, and the number of protons in the universe is

NEUTRINO ASTROPHYSICS

$$N_u = 10^{77} \text{ protons.} \tag{8.24}$$

Then for the universe, $r_s = 10^{25}$ cm. The radius of the universe is about 10^{28} cm. Hence, if all the matter of the universe is contained in galaxies, relativity is not important in cosmology. However, a large portion of the mass of the universe may be in the form of intergalactic matter without being detected.

Neutrinos are Fermi particles and therefore obey the exclusion principle. It has been estimated that if the universe were closed, so that neutrinos could not escape, then, at the death of the universe, the Fermi sea would be filled up by neutrinos to a depth of 500 eV.

At the Schwarzschild radius, g_{00}, the coefficient of dt^2, vanishes. The meaning of this is that light leaving a body enclosed within r_s would be infinitely red shifted. An observer outside r_s would say it takes an infinite time for light to reach him from inside. Thus, matter inside r_s is invisible to observers outside. Electrostatic or static gravitational fields of a Schwarzschild singularity can, however, still be detected. Thus a Schwarzschild singularity can be felt but not seen.

8.2 The Schwarzschild Interior Solution; The Equations of a Neutron Star

Inside the matter $T_{\mu\nu}$ is not equal to zero, but is given by

$$T = \begin{pmatrix} -P & & & \\ & -P & & \\ & & -P & \\ & & & \rho \end{pmatrix} ; \tag{8.25}$$

ρ is the energy density -- including matter.

In this case the gravitational equations are

i. $8\pi P = e^{-\lambda} \left(\dfrac{\nu'}{r} + \dfrac{1}{r^2}\right) - \dfrac{1}{r^2}$;

ii. $8\pi P = e^{-\lambda} \left(\dfrac{\nu''}{2} - \dfrac{\lambda'\nu'}{4} + \dfrac{\nu'^2}{4} + \dfrac{\nu' - \lambda'}{2r}\right)$; (8.26)

iii. $8\pi\rho = e^{-\lambda}\left(\dfrac{\lambda'}{r} - \dfrac{1}{r^2}\right) + \dfrac{1}{r^2}$.

To obtain an equation for the neutron star in a convenient form, we perform the following operations.

First, differentiate Eq. (8.26)(i):

$$8\pi \frac{dP}{dr} = e^{-\lambda}\left(\frac{\nu''}{r} - \frac{\nu'}{r^2} - \frac{2}{r^3}\right) - e^{-\lambda}\left(\frac{\lambda'\nu'}{r} + \frac{\lambda'}{r^2}\right) + \frac{2}{r^3} \qquad (8.27)$$

$$= \frac{2}{3}\left\{e^{-\lambda}\left(\frac{\nu''}{2} - \frac{\nu'}{2r} - \frac{\lambda'}{2r} - \frac{\lambda'\nu'}{2} - \frac{1}{r^2}\right) + \frac{1}{r^2}\right\}.$$

Next, subtract Eq. (8.26)(i) from Eq. (8.26)(ii) and multiply by $2/r$,

$$8\pi P - 8\pi P = 0 = \frac{2}{r}\left\{e^{-\lambda}\left(\frac{\nu''}{2} - \frac{\lambda'\nu'}{4} + \frac{\nu'^2}{4} - \frac{\nu'+\lambda'}{2r} - \frac{1}{r^2}\right) + \frac{1}{r^2}\right\}. \quad (8.28)$$

Substituting:

$$0 = 8\pi \frac{dP}{dr} + \left(\frac{\lambda'\nu'}{2} + \frac{\nu'^2}{2r}\right)e^{-\lambda}, \qquad (8.29)$$

$$0 = 8\pi\left(\frac{dP}{dr} + (P+\rho)\frac{\nu'}{2}\right).$$

This equation can be integrated:

$$\frac{d\nu}{dr} = -\frac{dP}{dr}\frac{2}{P+\rho}, \qquad (8.30)$$

$$\nu = \text{const} - \int_0^{P(r)} \frac{dP}{P+\rho(P)}.$$

Since $P = 0$ at $r = R$, the boundary of the star,

$$e^{\nu(r)} = e^{\nu(R)} \exp\left\{-2\int_0^{P(r)} \frac{dP}{P+\rho}\right\}. \qquad (8.31)$$

Outside the star, $T_{\mu\nu} = 0$ and the exterior solution should be used. This gives the boundary condition

$$e^{\nu(R)} = 1 - \frac{2m}{R} \quad . \tag{8.32}$$

This equation becomes, finally,

$$e^{\nu(r)} = (1 - \frac{2m}{R}) \exp\left\{-2 \int_0^{P(r)} \frac{dP}{P + \rho}\right\} . \tag{8.33}$$

To obtain a second equation, we define

$$u(r) = \frac{1}{2} r (1 - e^{-\lambda}) \quad . \tag{8.34}$$

Then

$$e^{-\lambda} = 1 - \frac{2u(r)}{r} \tag{8.35}$$

Clearly, since $e^{-\lambda(R)} = 1 - \frac{2m}{R}$,

$$u(R) = m , \tag{8.36}$$

and, from Eq. (8.34),

$$\frac{du(r)}{dr} = \frac{1}{2}(1 - e^{-\lambda}) + \frac{r}{2} \frac{d\lambda}{dr} e^{-\lambda} \tag{8.37}$$

$$= \frac{r^2}{2} \left\{ e^{-\lambda}(\lambda'/2 - 1/r^2) + 1/r^2 \right\}$$

$$= \frac{r^2}{2} \cdot 8\pi\rho = 4\pi r^2 \rho(r) .$$

From the equations

$$\frac{du}{dr} = 4\pi r^2 \rho , \quad u(R) = m , \tag{8.38}$$

it is obvious that $u(r)$ plays exactly the role of $m(r)$ in the non-relativistic calculation.

Using Eqs. (8.30) and (8.35), Eq. (8.26)(i) can be rewritten

$$8\pi P = (1 - \frac{2u}{r})[-\frac{1}{r}\frac{2}{P+\rho}\frac{dP}{dr} + \frac{1}{r^2}] - \frac{1}{r^2}, \qquad (8.39)$$

from which follows the relativistic form of the hydrostatic equilibrium equation,

$$\frac{dP}{dr} = -\frac{P+\rho}{r(r-2u)}\left\{4\pi P r^3 + u\right\}. \qquad (8.40)$$

Equations (8.38) and (8.40), together with an equation of state, are the basic equations of the neutron star. Recall the non-relativistic form,

$$\frac{dP}{dr} = -\frac{G\rho m}{r^2}.$$

Comparison reveals

$$\rho \rightarrow \rho + P,$$

$$m \rightarrow m + \text{const} \times P,$$

$$r^2 \rightarrow r^2 - 2mr.$$

The effect of the relativistic equations is to increase the gravitational force.
    Chandrasekhar, investigating the equilibrium of a star supported by the Fermi pressure on the electrons, found a mass greater than 1.44 $M_\odot$ could not be supported by the Fermi pressure, assuming the non-relativistic form of the hydrostatic equation. However, when the mass is around 1.2 $M_\odot$, β-decay already becomes important. Once neutrons are formed, the relatively high density ($\geq 10^{12}$ g/cm^3) causes an increase in the gravitational field from the contribution of the pressure term in the stress tensor. The mass limit further decreases to around 0.7 $M_\odot$.
    The central density as a function of mass is plotted in Fig. 8.1. The curve marked CH is a non-general-relativistic result of Chandrasekhar. The curve marked O-V is a general-relativistic result of Oppenheimer and Volkoff for an ideal Fermi gas. The solid curve is due to Wheeler <u>et al.</u>

NEUTRINO ASTROPHYSICS   85

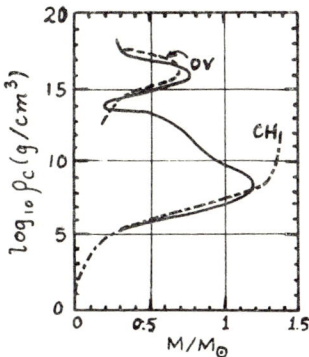

Fig. 8.1. Central Density vs. Mass for Zero-Temperature Stars. Solutions are stable where curve has positive slope, unstable where curve has negative slope.

8.3 The Equations of State

(a) <u>Fermi Gas</u>

The parametric form for the equation of state of matter composed of non-interacting Fermi particles at zero temperature is:

$$\rho = K(\sinh t - t),$$
$$P = \frac{1}{3}K(\sinh t - 8 \sinh \frac{1}{2} t + 3t), \qquad (8.41)$$
$$t = 4 \sinh^{-1} P_F/\mu_0 c;$$

P_F is the Fermi momentum and μ_0 the mass of the particles. K is a scale factor.

For a non-relativistic gas, $P_F \ll \mu_0 c$, $t \ll 1$. Then

$$\sinh t = t + t^3/3! + t^5/5! + \ldots$$

$$\rho = K(t + t^3/3! - t) = Kt^3/3!,$$

$$P = \frac{1}{3}K(t + t^3/3! + t^5/5! - 4t - t^3/3! - \frac{1}{4}t^5/5! - 3t)$$

$$= \frac{1}{3}K(\frac{3}{4}t^5/5!).$$

Therefore $P \propto \rho^{5/3}.$ (8.42)
This is the equation of state for a non-relativistic Fermi gas.

In the case $t \gg 1$, the relativistic case, the sinh t term dominates, so that

$$P = 1/3 \rho. \qquad (8.43)$$

Here ρ is the energy density.*

With this equation of state (Eq. (8.41)), the differential equations of the star become, by direct substitution, letting $K = 1/4\pi$,

$$\frac{du}{dr} = r^2 (\sinh t - t),$$

$$\frac{dt}{dr} = \frac{4}{r(r-2u)} \frac{\sinh t - 2 \sinh(t/2)}{\cosh t - 4 \cosh(t/2) + 3} \times \qquad (8.44)$$

$$[(\pi/3) r^3 (\sinh t - 8 \sinh(t/2) + 3t) + u].$$

The unit of length is now fixed:

$$r(\text{in cm}) = \frac{1}{\pi} (\frac{h}{m_n})^{3/2} \frac{c}{(m_n G)^{1/2}} r = (1.36 \times 10^6 \text{ cm}) r.$$

The unit of mass is fixed at 1.83×10^{34} gms.

The result of the calculation using this equation of state is given in the previous figure. If the structure is computed using a different equation of state, the details are changed but the qualitative result is the same. No stellar configurations exist for cold matter for masses greater than $\sim 1 M_\odot$. The central density is infinite at a finite mass.

(b) Hyperon Gas

Ambartsumian and Saakyan have investigated the conditions under which hyperons and μ and π mesons would be formed in a dense body, by the reactions,

---

*In the case of the ordinary relativistic Fermi gas, we frequently write $P \propto \rho^{4/3}$ where ρ in this case is the matter density (see Landau and Lifshitz, <u>Statistical Physics</u>, p. 166). This has been a source of confusion in the literature.

$$n \rightleftharpoons P + e^- + \bar{\nu} \uparrow$$
$$\Lambda \rightleftharpoons P + e^- + \bar{\nu} \uparrow$$
$$\Sigma^° \rightleftharpoons P + e^- + \bar{\nu} \uparrow$$
$$\Sigma^- \rightleftharpoons n + e^- + \bar{\nu} \uparrow$$
$$\Xi^- \rightleftharpoons \Lambda + e^- + \bar{\nu} \uparrow$$
$$\Xi^° \rightleftharpoons \Sigma + e^- + \bar{\nu} \uparrow$$
$$\pi \rightleftharpoons \mu + \bar{\nu}_\mu \uparrow$$

The neutrinos formed are each time, in this model, allowed to escape from the star.

When the μ Fermi sea is filled to the level of the π mass, pions will not be able to decay, and they will form a degenerate Bose gas.

A summary of the hyperonic properties of the star is summarized below:

ρ

$< 1.28 \times 10^7$ g/cm^3 — ordinary matter from He^4 to heavy elements.

$= 1.28 \times 10^7$ g/cm^3 — threshold for inverse beta decay.

3×10^{11} g/cm^3 — threshold for a neutron gas.

1.1×10^{15} g/cm^3 — threshold for hyperon formation First to appear is Σ^-.

2.36×10^{15} g/cm^3 — Λ production threshold.

$\sim 10^{16}$ g/cm^3 — All baryons above present to the same order of magnitude.

$> 5 \times 10^{16}$ g/cm^3 — The situation is vague. Nuclear repulsive forces would begin to be important. Higher baryon resonances would occur. π-mesons might be formed.

Using an equation of state based on the above analysis of hyperon formation, Ambartsumian and Saakyan also calculated stellar configurations. They obtained a similar result to that of Oppenheimer and Volkoff.

(c) Incompressible Fluid

It is not possible to have in nature an incompressible fluid, because then the speed of sound would exceed the speed of light. The speed of sound is

$$v_s = c\sqrt{\left(\frac{dP}{d\epsilon}\right)_{ad}} \quad . \tag{8.45}$$

P is the pressure and E the energy density.

For an incompressible fluid v_s is infinite.
For light $(1/3)E = P$, so that

$$v_s = (1/\sqrt{3})c \quad .$$

The stress tensor, Eq. (8.25), Eq. (8.25), with the energy density now written ϵ,

$$T_{\mu\nu} = \begin{pmatrix} -P & & & \\ & -P & & \\ & & -P & \\ & & & \epsilon \end{pmatrix} \quad ,$$

must be positive for an ideal, non-interacting gas. This means the trace must be positive, so

$$-3P + \epsilon > 0 \quad ,$$

$$P < 1/3\epsilon \quad .$$

This relation, which limits the pressure a gas can have, was believed to hold quite generally, although the limit obtained by integrating the expression $\quad v_s = c\sqrt{\frac{dP}{d\epsilon}} \leq c \quad$ is

$$P \leq \epsilon \quad . \tag{8.46}$$

That the limit $P \leq \epsilon$, rather than the more restrictive $P \leq 1/3\epsilon$ can actually be obtained, was discovered by Zel'dovich. His model is a classical vector field with mass, interacting with stationary classical point charges. In this model

$$\epsilon = Mn + 2\pi g^2 n^2/\mu^2 \quad ,$$
$$P = 2\pi g^2 n^2/\mu^2 \quad , \tag{8.47}$$

NEUTRINO ASTROPHYSICS   89

where $-g^2 \frac{d}{dr_{12}}\left[e^{-\mu r_{12}}/r_{12}\right]$ is the repulsive force between charges, M is their mass and n is the number density. In the limit of large density $\rho = \epsilon \propto n^2$. Also $v_s = c$.

Such an equation of state is the most rigid permissible by relativity. Nevertheless, it is interesting to consider the unphysical case of an incompressible fluid to see if in this extreme case a massive neutron star can exist. This investigation was undertaken by Wheeler.

For an incompressible fluid, ρ = const.

Using this equation, (8.26)(iii) above can be rewritten and integrated

$$8\pi\rho = -\frac{e^{-\lambda}}{r^2}[-\lambda' r + 1] + 1/r^2 \qquad (8.48)$$

$$= -\frac{1}{r^2}\frac{d}{dr}[re^{-\lambda}] + 1/r^2 .$$

Then

$$e^{-\lambda} = 1 - \frac{8\pi\rho}{3} r^2 + C/r^2 . \qquad (8.49)$$

If there is not to be a singularity at the origin C must be zero. The result is conveniently written

$$e^{-\lambda} = 1 - (r/R)^2 , \qquad R = \sqrt{3/8\pi\rho} . \qquad (8.50)$$

The relation (8.30),

$$\frac{dP}{dr} = (P+\rho)\frac{\nu'}{2},$$

can now also be integrated and gives

$$P + \rho = \text{const } e^{-\nu/2} . \qquad (8.51)$$

Adding Eqs. (8.26)(i) and (iii),

$$8\pi(P+\rho) = e^{-\lambda}(\nu' + \lambda')/r .$$

Combining these equations, we find

$$e^{\nu/2}(2/R^2 + \nu'/r - r\nu'/R^2) = \text{const.} \qquad (8.52)$$

The solution is

$$e^{\nu/2} = A - B\sqrt{1 - r^2/R^2}. \qquad (8.53)$$

Therefore the metric inside a medium of incompressible fluid is

$$ds^2 = (A - B\sqrt{1 - r^2/R^2})^2 dt^2 - \frac{dr^2}{1 - r^2/R^2} - r^2 d\theta^2 - r^2 \sin^2\theta d\phi^2. \quad (8.54)$$

Substitution of the above equations (8.50) and (8.53) into Eq. (8.26)(i) gives

$$8\pi P = \frac{1}{R^2} \left[\frac{3B\sqrt{1 - r^2/R^2} - A}{A - B\sqrt{1 - r^2/R^2}} \right]. \qquad (8.55)$$

For the metric above to join onto the Schwarzschild exterior solution

$$(A - B\sqrt{1 - r_1^2/R^2})^2 = 1 - 2m/r_1 = 1 - r_1^2/R^2, \qquad (8.56)$$

where $R^2 = 3/8\pi\rho$ and r_1 is the stellar boundary.

This determines the constants

$$B = 1/2,$$

$$m = (4\pi/3)\rho r_1^3,$$

$$A = \frac{3}{2}\sqrt{1 - r_1^2/R^2},$$

and with this result the interior metric becomes

$$ds^2 = \frac{1}{4}(3\sqrt{1 - r_1^2/R^2} - \sqrt{1 - r^2/R^2})^2 \, dt^2 -$$
$$- [\frac{dr^2}{1 - r^2/R^2} + r^2 d\theta^2 + r^2 \sin^2\theta \, d\phi^2] \, ; \quad (8.58)$$

$g_{oo} = e^{\nu/2}$ measures the energy separation between Fermi particles of positive and negative energy.

$$g_{oo} \geqslant 0 \, .$$

The limit $g_{oo} = 0$ can be reached at $r = 0$ for the finite stellar radius r_1:

$$g_{oo}(r = 0) = \frac{1}{2}(3\sqrt{1 - r_1^2/R^2} - 1) \, ; \quad (8.59)$$

$g_{oo} = 0$ when

$$r_1/R = \sqrt{1 - (1/3)^2} = \sqrt{8/9} \, . \quad (8.60)$$

At this radius, particle pairs can be created with no energy at the center of the star. Photons created at the center will be infinitely red shifted and have zero energy when they come out.

This corresponds to infinite pressure at the origin, for the pressure has a singularity as $r \to 0$:

$$8\pi P = \frac{3[1 - (r/R)^2]^{1/2} - 3[1 - r_1^2/R^2]^{1/2}}{3[1 - r_1^2/R^2]^{1/2} - [1 - r^2/R^2]^{1/2}} \, . \quad (8.61)$$

At $r_1/R = \sqrt{8/9}$,

$$8\pi P = \frac{3(1 - [r/R]^2)^{1/2} - 1}{1 - (1 - r^2/R^2)^{1/2}} \sim \frac{1}{r} \xrightarrow[r \to 0]{} \infty \, .$$

The mass of the system, before it was put together as a star under gravitational attraction, **was finite**.

$$M = \int (\rho + 3P)(r\sin\theta\, d\phi)(r\, d\theta)(e^{1/2\lambda(r)}\, dr). \qquad (8.62)$$

The conclusion that this leads us to is that the existence of a maximum mass is not prevented by a sufficiently rigid equation of state.

We have found here no way of constructing a model for a stable star of larger than a few solar masses.

8.4 Discussion

What happens in Wheeler's model, if some mass is added above the maximum mass? The result suggested by Wheeler is that the mass is converted entirely into radiation -- gravitational, electromagnetic or neutrino. This would violate the law of conservation of baryon number. The present lower limit on the lifetime of a nucleon is $T_{1/2} > 4 \times 10^{23}$ years, but this measurement by Cowan and Reines is based on normal pressures and values of g_{oo} very close to 1.

Another possibility is that no stable configuration exists, that for masses above the critical mass the star will simply collapse. Oppenheimer and Snyder computed the general characteristics of a star collapsing with zero pressure to within its Schwarzschild radius. An observer on the star would say the collapse took a finite time. For the sun it would be about a day. But to a distant observer, due to the time dilation as g_{oo} increases, the time of fall would be infinite.

This model is somewhat unrealistic since the pressure cannot be zero but must be $P \sim \rho c^2$. It remains possible that in this case also g_{oo} would become infinite at the center in a finite time.

Another possibility is the existence of a super strong interaction with a range much smaller than that of the short range repulsive force. The range must be small enough to explain why it has not been observed. This attractive force must cause a local clustering of particles into quasi bound states. The energy of the system will be decreased by the amount of the binding energy, and consequently the mass limit can be increased.

If all the mass of the universe is included in a neutron star, to save the law of baryon conservation the super-strong interaction would have to have the following properties: The coupling constant, $f^2/\hbar c$, must be 100 times stronger than that of the strong interactions. It must be a scalar or tensor force (so that the force is attractive between all pairs of particles). The range must be $< 10^{-16}$ cm since there is no evidence for it. The quanta must be about 100 times the π meson mass.

NEUTRINO ASTROPHYSICS

There are many objects in the universe with masses greater than a few solar masses. Therefore, it is a real, as well as theoretically interesting, problem to inquire what is the final state of systems of large mass after all energy generation processes have ceased. For a stable configuration to exist, it seems that some radical change in our understanding of elementary particles may be necessary. The analysis of unstable collapsing states is still incomplete. At the moment our conclusions remain speculation.

8.5 Observation of Neutron Stars

Objects with masses less than the critical mass, formed from large objects which eject most of their mass in a supernova explosion will form neutron stars rather than white dwarfs. For a star of mass M, radius R, to eject a mass $\approx$ M requires an energy

$$E \sim \frac{GM^2}{R},$$

which must come from the binding energy of the new configuration with mass m and radius r:

$$E_B \sim \frac{Gm^2}{r}.$$

Then

$$\frac{GM^2}{R} \sim \frac{Gm^2}{r}.$$

For $M \sim 20\, M_\odot$, $m \sim M_\odot$ and $R \sim 10^9$ cm, we find $r \sim R/400$, and this corresponds to a density $\rho \sim 10^{14}$ g/cc, the density of neutron stars.

How would one observe such a star? It would have the structure illustrated below:

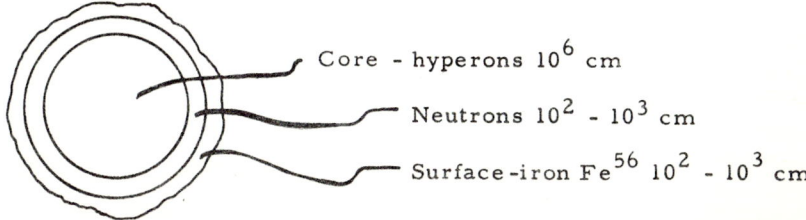

Core - hyperons 10^6 cm

Neutrons $10^2 - 10^3$ cm

Surface-iron Fe^{56} $10^2 - 10^3$ cm

In the early period of its life it would have these properties:

$$\text{Energy content} \sim E_\odot \sim 4 \times 10^{47} \text{ erg,}$$

$$\text{Luminosity} \sim 2 \times 10^3 \ L_\odot \sim 10^{37} \text{ ergs/sec,}$$

$$\text{Temperature} \sim 10^7 \ ^\circ K.$$

It would last about 1000 years in this stage before cooling to $T \sim 10^{6} \ ^\circ K$ with $L \sim 10^{33}$ erg/sec, when it gradually becomes unobservable.

Our atmosphere is transparent to radiation only in the following windows: 3000 A - 7000 A, visible; 8000 A - 12,000 A, infrared; 1 cm - 10 meters, microwave; and 200 m - ∞, low frequency radio-waves.

Most radiation from neutron stars would have a wavelength $\sim$ 10 A, not in one of the windows. The apparent visible luminosity of neutron stars would be $10^{-8} \ L_\odot$.

There are well-known techniques for observing light in certain frequency ranges -- γ-rays by Geiger counters; 1 A by proportional counters; 100 A - 3000 A by photoelectric devices. There is no such technique for 10 A radiation, although some solid state devices show promise. The development of such technology, however, together with the possibility of mounting it in an earth satellite, would make neutron stars detectable.

I know of no suggestions for observing the Schwarzschild singularities.

P.S. At the present time H. Friedman (to be published in Nature) and his group, using rockets, have detected certain x-ray sources which are possibly associated with supernova remnants 1054 A.D. (Chinese observation) and 827 A.D. (Arabic observation), with the same characteristics in total energy output and wavelength region predicted here.

9. A WORD ABOUT COSMOLOGY

In the 17th century Messier made a catalog of 103 objects in the sky. Included among them were several that were extragalactic. M1 was the crab nebula, M31 Andromeda, M87.
The first suggestion that these objects were extragalactic was made in 1905 by Shapley. The discovery of Miss Leavitt that Cepheid variables could be used to measure distances of distant objects was a major step. In 1920, Hubble related the red shift of light from distant objects to distance measurements.

Different world models present different observational features. The 200" telescope was built with the hope of obtaining accurate measurements of galactic density to distinguish between world models. But the discovery of an error in the calibration of distance measurements and the subsequent re-evaluation of all distances in 1952 pushed the resolution necessary beyond the capability of the telescope.

For example, Fig. 9.1 gives the red shift as a function of magnitude for different models. The circle is at the limit of the telescope and indicates the resolution.

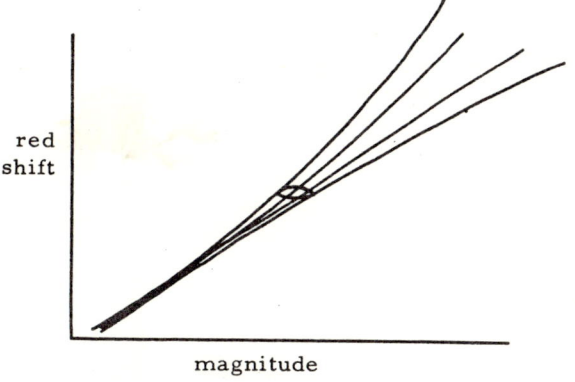

Fig. 9.1.

If galaxies are distributed uniformly in space, the number enclosed within the distance r will be proportional to the volume. The volume of curved space will vary slower or faster than r^3 depending on whether space is closed or open (equals r^3 for Euclidean).

The difference between models in this case, at the telescope limit, however, is one-third as large as in the previous case.

A third possible test is the angular diameter of distant objects. To see how this works, consider three equal-sized objects on a plane and on the surface of a sphere.

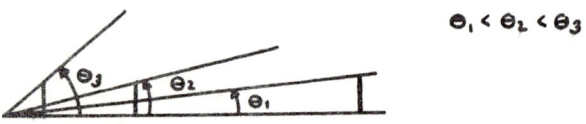

In the plane geodesics are straight lines and the angular diameter of similar objects decreases as a function of distance.

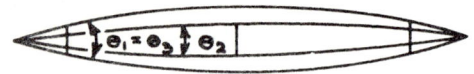

On the sphere geodesics are great circles and the angular diameter as a function of distance first decreases, then increases again. The diagram is again similar.

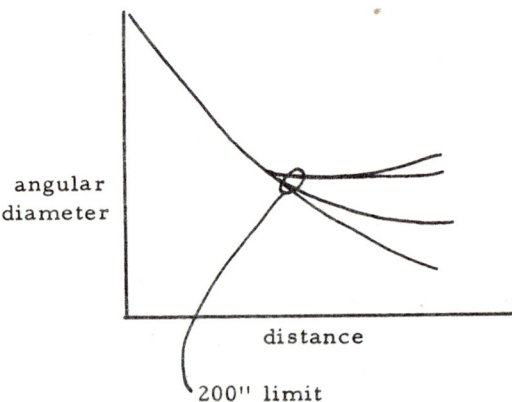

In all cases the 200" telescope has nearly, but not quite, the ability to distinguish between different world models.

10. SUMMARY

I want to summarize the physics that is appropriate to the different stellar situations that we have discussed.

Mechanics

The stability of a star is guaranteed by the equations
$E_T < 0$, $\delta E_T = 0$, $\delta^2 E_T < 0$.

One can then, in the classical way, determine stellar mechanics -- which largely determines the structure of the star.

- sun — Newtonian mechanics, hydrostatics

- red giant, white dwarf } special relativity but classical gravitation

- neutron stars, Schwarzschild singularity } special relativity, general relativity (?)

Evolution

Evolution is governed by dE_r/dt, the energy flow out of a star. Gravitational energy is released and kinetic energy increases.

- $T = 300\ °K$ ordinary solid
- $10^4\ °K$ ionized gas
- $10^7\ °K$ nuclear reactions - energy production
- $10^9\ °K$ nuclear reactions become endothermal; neutrinos; stellar collapse
- $10^{10}\ °K$ could not exist at this temperature very long

$\rho = 1\text{-}10$ g/cc ordinary matter

10^{-8} g/cc, 10^{4o}K laboratory plasma

1 g/cc, $10^4\text{-}10^7$ °K ordinary plasma - sun

$>10^4$ g/cc degenerate plasma

10^{14} g/cc neutrons, hyperons nuclear matter

No states can exist with $\rho = 10^8\text{-}10^{13}$ g/cc and $kT \ll E_F$ because of inverse β-decay.

Element Synthesis

1. The first idea was that elements are all created in the beginning.
2. Elements built up in stars.
   a) First, it was not seen how to go beyond the light elements.
   b) Carbon generation and α process and heavier elements, but still not all elements produced.
   c) Equilibrium process--iron group heavy elements in supernova explosion.
   d) Fast, slow, neutron capture--elements heavier than iron. Reasonably satisfactory numerical agreement with element abundance.

On stellar surfaces atomic physics is important. The stellar interior is an application of nuclear physics. Late stages of stellar evolution involve elementary particle physics.

APPENDIX

A. Rotation

The rotation of a sphere will make it bulge from the center.

$\omega = 0$ $\qquad\qquad\omega > 0$

In order that a rotating sphere be gravitationally bound, the gravitational force per unit mass at a radius must be greater than the centrifugal force:

$$\frac{v^2}{R} \leq \frac{GM}{R^2} \qquad (A.1)$$

$$v = \omega R$$

$$\therefore \quad R^3 \omega^2 \leq GM ,$$

$$\omega^2 \leq \frac{GM}{R^3} \text{ (for the sun)} = \frac{(7 \times 10^{-8})(2 \times 10^{33})}{(7 \times 10^{10})^3} \qquad (A.2)$$

$$= 40 \times 10^{-8} ,$$

$\omega < 6.3 \times 10^{-4}$ rad/sec for the sun to be rotationally stable. For the solar rotation period,

$$\omega = 2\pi/T = 2\pi/26 \times 10^5 = 2.5 \times 10^{-6} \text{ rad/sec},$$

$$\omega/\omega_{max} \approx 1/50 .$$

The solar rotation frequency is not very much smaller than the maximum frequency at which the sun could rotate and remain gravitationally bound.

Solar rotation may have consequences for the test of general relativity by the perihelion of mercury.

Brouwer has expanded the gravitational potential of a non-symmetric object:

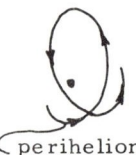

$$V = V(r) + P_2(\cos \theta) V_2(r) + \ldots$$

axis of rotation

He found that a body orbiting such a sphere would find its perihelion rotating at a frequency

$$\omega' = \frac{6\pi}{5} \epsilon \left(\frac{R}{r}\right)^2 \times (\text{geometrical factor}) \ . \quad (A.3)$$

The perihelion is the point on the orbit which is closest to the attraction center.

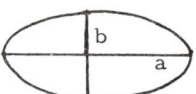

Rotation of the orbit.

perihelion

ϵ is the eccentricity of the central body defined as

$$\epsilon = \frac{b-a}{a} \ ;$$

a is the major axis and b is the minor axis.

In the table below we list the factors that contribute to the rotation of the perihelion of mercury and, for contrast, the earth.

There is a fairly large discrepancy in the perihelion advance of Mercury. Einstein's theory of general relativity predicts an advance in addition to those given above, of

$$\frac{6\pi GM}{R_{\text{planet}}^2 \, R_\odot^2} \, \frac{1}{1-e^2} = 43''.03 \pm 0.01 \ ; \quad (A.4)$$
orbit

e is the eccentricity of the planetary orbit.

NEUTRINO ASTROPHYSICS 101

Contribution from	Mercury	Earth
Mercury	0.025 ± 0.00*	
Venus	277.856 ± 0.68	345.49 ± 0.8
Earth	90.038 ± 0.08	-
Mars	2.536 ± 0.00	97.69 ± 0.1
Jupiter	153.584 ± 0.00	696.85 ± 0.1
Saturn	7.302 ± 0.01	18.74
Uranus	0.141 ± 0.00	0.57
Neptune	0.042 ± 0.00	0.18
Moon	-	7.68
general precession of the equinox	5025.645 ± 0.5	5025.65
Total	5557.1 ± 0.8	6179.1 ± 2.5
Experimentally observed	5599.7 ± 0.4	6183.7 ± 1.1
Difference	42.6 ± 0.9	4.6 ± 2.7

*This is due to relative orbital motion of Mercury and Earth.

The gravitational perihelion advance may be written $\omega_g = 6\pi \times 2 \times 10^{-8}$, while the advance calculated from Eq. (A.3), due to the sun being aspherical, is $\omega_a = 6\pi \times 6 \times 10^{-4} \epsilon$.

If we believe in general relativity, then $\epsilon \ll 10^{-4}$ despite the relatively high value of the rotational frequency.

The possibility of using Venus as an independent check is doubtful. Mercury's orbital elements are measured during its transits of the sun. Venus has far fewer transits. And none this century.

The non-sphericity of the earth itself is quite large, and consequently it is hopeless to check general relativity by measuring the perihelion advance of earth satellites. However, it is possible to test the Lense-Thirring effect (precession of a vector (the angular momentum of a gyroscope) moving in the field of the earth) by a drag-free satellite.

B. Determination of the Temperature of a Star

The energy spectrum is

$$E(\nu)d\nu \propto \frac{\nu^3 \, d\nu}{e^{\nu/kT} - 1}.$$

In the two-color method a reading is taken of the intensity of radiation at two different frequencies. The temperature can be determined by a comparison of the two intensities. In practice, filters are used to select the frequencies. Filters are not monochromatic and accept a range of frequencies. The intensity E_i accepted by the filters may be written

$$E_1 = \int_0^\infty f_1(\nu) E(\nu) \, d\nu,$$

$$E_2 = \int_0^\infty f_2(\nu) E(\nu) \, d\nu.$$

Somewhat idealized filters might have

$$f_1(\nu) = 0, \qquad \nu \geqslant \nu_1;$$
$$f_1(\nu) = 1, \qquad \nu \leqslant \nu_1;$$

$$f_2(\nu) = 0, \qquad \nu \leqslant \nu_2;$$
$$f_2(\nu) = 1, \qquad \nu \geqslant \nu_2;$$

then

$$E_1(\nu_1, T) = \int_0^{\nu_1} E(\nu) d\nu,$$
$$E_2(\nu_2, T) = \int_{\nu_2}^\infty E(\nu) d\nu.$$

The ratio of E_1 to E_2 can be calibrated to read temperature, although often results are plotted in terms of the B-V color index directly without calibration.

This method suffers from the fact that stars are not exactly black bodies. In particular, the spectra have absorption lines, and some strong lines in a star may alter the calibration of the temperature measure for that star. Sometimes a three-color index is used. This has the advantage of being able to identify certain spectral features and consequently to eliminate their effects from the temperature calibration.

BIBLIOGRAPHY

Chapter 2

A general reference is:

M. Schwarzschild, Stellar Structure and Evolution (Princeton Univ. Press, 1958).

On the mass limit for white dwarfs:

S. Chandrasekhar, "Stellar Configurations and White Dwarfs," Ch XI, Stellar Structure (Dover, 1957 (original copyright, 1939)).

For the energy formula for degenerate gas see:

J. A. Wheeler, "Superdense Stars," Ch. X, Relativity and Gravitation, ed. H.-Y. Chiu and W. F. Hoffman (W. A. Benjamin, 1963).

Our formulation of the equilibrium condition is due to:

F. J. Dyson, "Hydrostatic Instability of a Star," unpublished, Ch. III.

For discussion of the equation of radiative transfer see:

S. Chandrasekhar, "The Equation of Transfer" in Radiative Transfer (Dover, 1960).

On opacity see:

M. Schwarzschild, section 9.

Opacity for the Russell mixture is given in:

S. Chandrasekhar, Stellar Structure, p. 255 ff.

Chapter 4

A general discussion of nuclear reaction theory is given by:

J. M. Blatt and V. F. Weisskopf, Theoretical Nuclear Physics (John Wiley and Sons, 1952).

A review paper on nuclear reactions in astrophysics is:

H. Reeves, Stellar Energy Sources, NASA Institute for Space Studies, preprint.

See also:

M. Schwarzschild, section 10.

On beta decay:

R. B. Leighton, Principles of Modern Physics (McGraw-Hill, 1959).

For pycnonuclear reactions see:

A. G. W. Cameron, "Pycnonuclear Reaction and Nova Explosions," Astrophys. J. **130**, 916 (1959).

E. E. Salpeter, "Energy and Pressure of a Zero Temperature Plasma," Astrophys. J. **134**, 669 (1961).

For the Wigner-Seitz method:

E. Wigner and F. Seitz, "On the Constitution of Metallic Sodium," Phys. Rev. **43**, 804 (1933).

C. Kittel, Introduction to Solid State Physics (John Wiley and Sons, 1953), p. 285.

Chapter 5

C. Hayashi, R. Hashi and D. Sugimoto, "Evolution of the Stars," Progress of Theoretical Physics, Supp. No. 22, 1962.

"Michelson's Stellar Interferometer," section 16.8, in F. A. Jenkins and H. E. White, Fundamentals of Optics (McGraw-Hill, 1957).

Chapter 6

For URCA process:

G. Gamow and M. Schönberg, Phys. Rev. **59**, 539 (1941).

H.-Y. Chiu, Ann. Phys. **15**, 1 (1961).

For neutrino astrophysics see:

J. B. Adams, M. A. Ruderman, and C. H. Woo, "Neutrino Pair Emission by a Stellar Plasma," Phys. Rev. 129, 1383 (1963).

H.-Y. Chiu and R. C. Stabler, "Emission of Photoneutrinos and Pair Annihilation from Stars," Phys. Rev. 122, 1317 (1961), and Phys. Rev. 131, 2839 (1963).

H.-Y. Chiu, Ann. Phys. 16, 321 (1961).

V. I. Ritus, "Photoproduction of Neutrinos on Electrons and Neutrino Radiation from Stars," Soviet Phys.--JETP 14, 915 (1962).

Also:

C. Hayashi and A. G. W. Cameron, "The Evolution of Massive Stars, Astrophys. J. 136, 166 (1962).

A. G. W. Cameron, "Neutron Star Models," Astrophys. J. 130, 884 (1959). This paper contains an error. See G. S. Saakyan, Astro. Zhur. 40, 82 (1963), translation Soviet Astron.--AJ 7, 60 (1963).

Chapter 7

General references are, for example:

C. Møller, The Theory of Relativity, (Oxford Univ. Press, 1952).

R. C. Tolman, Relativity, Thermodynamics and Cosmology (Oxford Univ. Press, 1934).

See also:

Relativity and Gravitation, ed. H.-Y. Chiu and W. F. Hoffman, especially the Introduction.

E. P. Wigner, "Relativistic Invariance and Quantum Phenomena," Revs. Modern Phys. 29, 255 (1957).

E. Wigner and H. Salecka, "Quantum Limitations of the Measurements of Space-Time Distances," Phys. Rev. 109, 571 (1958).

Chapter 8

V. A. Ambartsumian and G. S. Saakyan, "The Degenerate Superdense Gas of Elementary Particles," Soviet Astron. --AJ 4, 187-354 (1960).

V. A. Ambartsumian and G. S. Saakyan, "On Equilibrium Configurations of Superdense Degenerate Gas Masses," Soviet Astron. --AJ 5, 701 (1962).

H.-Y. Chiu, "Supernovae, Neutrinos and Neutron Stars," NASA Institute for Space Studies, preprint (to be published in Ann. Phys.).

H.-Y. Chiu, "Two Neutrinos and Neutron Stars," NASA Institute for Space Studies, preprint 7.

J. R. Oppenheimer and R. Serber, "On the Stability of Stellar Neutron Cores," Phys. Rev. 54, 540 (1938).

J. R. Oppenheimer and H. Snyder, "On Continued Gravitational Contraction," Phys. Rev. 56, 455 (1939).

J. R. Oppenheimer and G. Volkoff, "On Massive Neutron Cores," Phys. Rev. 55, 3741 (1939).

G. Volkoff, "On the Equilibrium of Massive Spheres," Phys. Rev. 55, 413 (1939).

G. S. Saakyan, "On the Non-Relativistic Theory of Superdense Stellar Configurations," Soviet Astron. --AJ 6, 783 (1963).

R. Tolman, "Static Solutions of Einstein's Field Equations for Spheres of Fluid," Phys. Rev. 55, 364 (1939).

J. A. Wheeler, Chapter 10 in *Relativity and Gravitation*.

A. B. Zel'dovich, "The Equation of State at Ultrahigh Densities and Its Relativistic Limitations," Soviet Phys. --JETP 14, 1143 (1962).

Chapter 9

A. R. Sandage, "The Ability of the 200-Inch Telescope to Discriminate between Selected World Models," Soviet Astron. --AJ 133, 355 (1961).

Appendix

G. M. Clemence, "The Relativity Effect in Planetary Motions,"
   Revs. Modern Phys. 19, 361 (1947).